Hilbert Transform Applications in Mechanical Vibration

Hilbert Transform Applications in Mechanical Vibration

Michael Feldman

Technion - Israel Institute of Technology, Israel

A John Wiley and Sons, Ltd., Publication

This edition first published 2011
© 2011 John Wiley & Sons, Ltd.

Registered office
John Wiley & Sons Ltd, The Atrium, Southern Gate, Chichester, West Sussex, PO19 8SQ, United Kingdom

For details of our global editorial offices, for customer services and for information about how to apply for
permission to reuse the copyright material in this book please see our website at www.wiley.com.

Library of Congress Cataloging-in-Publication Data

Feldman, Michael, 1951-
    Hilbert transform applications in mechanical vibration / Dr. Michael Feldman.
      p. cm.
    Includes bibliographical references and index.
  ISBN 978-0-470-97827-6 (hardback)
  1. Vibration–Mathematical models.   2. Hilbert transform.   I. Title.
  TA355.F35 2011
  620.301′515723–dc22

                                                                        2010051079

A catalogue record for this book is available from the British Library.

Print ISBN: 9780470978276 (H/B)
E-PDF ISBN 9781119991649
O-book ISBN 9781119991656
E-Pub ISBN 9781119991526

Typeset in 10/12pt Times by Aptara Inc., New Delhi, India

Contents

PART II   HILBERT TRANSFORM AND VIBRATION SIGNALS

List of Figures

List of Tables

Preface

The object of this book, *Hilbert Transform Applications in Mechanical Vibration*, is to present a modern methodology and examples of nonstationary vibration signal analysis and nonlinear mechanical system identification. Nowadays the Hilbert transform (HT) and the related concept of an analytic signal, in combination with other time–frequency methods, has been widely adopted for diverse applications of signal and system processing.

What makes the HT so unique and so attractive?

- It solves a typical demodulation problem, giving the amplitude (envelope) and instantaneous frequency of a measured signal. The instantaneous amplitude and frequency functions are complementary characteristics that can be used to measure and detect local and global features of the signal – in the same way as for classical spectral and statistical signatures.

- The HT allows us to decompose a nonstationary complicated vibration, separating it into elementary time-varying components – preserving their shape, amplitude, and phase relations.

- It identifies and has an ability to capture – in a much faster and more precise way – the dynamic characteristics of system stiffness and damping, including their nonlinearities and the temporal evolution of modal parameters. This allows the development of more adequate mathematical models of tested vibration structures.

The information obtained can be further used in design and manufacturing to improve the dynamic behavior of the construction, to plan control actions, to instill situational awareness, and to enable health monitoring and preventive surplus maintenance procedures. Therefore, the HT is very useful for mechanical engineering applications where many types of nonlinear modeling and nonstationary parametric problems exist.

This book covers modern advances in the application of the Hilbert transform in vibration engineering, where researchers can now produce laboratory dynamic tests more quickly and accurately. It integrates important pioneering developments of signal processing and mathematical models with typical properties of mechanical construction, such as resonance, dynamic stiffness, and damping. The unique merger of technical properties and digital signal processing provides an instant solution to a variety of engineering problems, and an in-depth exploration of the physics of vibration by analysis, identification, and simulation. These modern methods of diagnostics and health monitoring permit a much faster development, improvement, and economical maintenance of mechanical and electromechanical equipment.

The Hilbert Vibration Decomposition (HVD), FREEVIB, FORCEVIB, and congruent envelope methods presented allow faster and simpler solutions for problems – of a high-order and at earlier engineering levels – than traditional textbook approaches. This book can inspire further development in the field of nonlinear vibration analysis with the use of the HT.

Naturally, it is focused only on applying the HT and the analytic signal methods to mechanical vibration analysis, where they have greatest use. This is a particular one-dimensional version of the application of HT, which provides a set of tools for understanding and working with a complex notation. HT methods are also widely used in other disciplines of applied mechanics, such as the HT spectroscopy that measures high-frequency emission spectra. However, the HT is also widely used in the bidimensional (2D) case that occurs in image analysis. For example, the HT wideband radar provides the bandwidth and dynamic range needed for high-resolution images. The 2D HT allows the calculation of analytic images with a better edge and envelope detection because it has a longer impulse response that helps to reduce the effects of noise.

HT theory and realizations are continually evolving, bringing new challenges and attractive options. The author has been working on applications of the HT to vibration analysis for more than 25 years, and this book represents the results and achievements of many years of research. During the last decade, interest in the topic of the HT has been progressively rising, as evidenced by the growing number of papers on this topic published in journals and conference proceedings. For that reason the author is convinced that the interest of potential readers will reach its peak in 2011, and that this is the right time to publish the book.

The author believes that this book will be of interest to professionals and students dealing not only with mechanical, aerospace, and civil engineering, but also with naval architecture, biomechanics, robotics, and mechatronics. For students of engineering at both undergraduate and graduate levels, it can serve as a useful study guide and a powerful learning aid in many courses such as signal processing, mechanical vibration, structural dynamics, and structural health monitoring. For instructors, it offers an easy and efficient approach to a curriculum development and teaching innovations.

The author would like to express his utmost gratitude to Prof. Yakov Ben-Haim (Technion), Prof. Simon Braun (Technion), and Prof. Keith Worden (University of Sheffield) for their long-standing interest and permanent support of the research developments included in this book.

The author has also greatly benefited from many stimulating discussions with his colleagues from the Mechanical Engineering Faculty (Technion): Prof. Izhak Bucher, Prof. David Elata, Prof. Oleg Gendelman, and Prof. Oded Gottlieb. These discussions provided the thrust for the author's work and induced him to continue research activities on the subject of Hilbert transforms.

The book summarizes and supplements the author's investigations that have been published in various scientific journals. It also reviews and extends the author's recent publications: Feldman, M. (2009) "Hilbert transform, envelope, instantaneous phase, and frequency", in *Encyclopedia of Structural Health Monitoring* (chapter 25). John Wiley & Sons Ltd; and Feldman, M. (2011) "Hilbert transform in vibration analysis" (tutorial), *Mechanical Systems and Signal Processing*, **25** (3).

The author is very grateful to Donna Bossin and Irina Vatman who had such a difficult time reading, editing, and revising the text. Of course, any errors that remain are solely the responsibility of the author.

Michael Feldman

1

Introduction

In contrast with other integral transforms, such as Fourier or Laplace, the Hilbert transform (HT) is not a transform between domains. It rather assigns a complementary imaginary part to a given real part, or vice versa, by shifting each component of the signal by a quarter of a period. Thus, the HT pair provides a method for determining the instantaneous amplitude and the instantaneous frequency of a signal. Creating and applying such complementary component seems to be a simple task. Nevertheless providing explanations and justifying the HT application in vibration analysis is a rather uneasy mission. There are a number of objective reasons complicating the matter.

First, the HT mathematical definition itself was originated just 60 years ago – not as long ago as the Fourier transform (Therrien, 2002), for example. Even 30 years ago the HT was a pure theory, and then it was employed in applied researches, including vibration. Thus the most significant and interesting results have been received within the last 15 years. Presenting this material, together with the corresponding statements and judgments, requires considerable efforts.

Secondly, from the very beginning the HT approach faced numerous objections, doubts, counterexamples, paradoxes, and alternatives generating some uncertainty about the reliability and feasibility of the obtained results. Naturally, the book should contain only proved, tested, and significant results of HT applications.

Thirdly, at the same time, and in parallel with the HT, another method – the Wavelets transform – was developed in signal processing allowing us to solve similar applied problems. As numerous scientific works devoted to the evaluation of these methods are based exclusively on a comparison of empirical data, theoretical conclusions and statements have not yet been available for detailed presentation.

Fourthly, the HT itself, and the corresponding methods of signal processing, involve rather difficult theoretical and empirical constructions, while the text should be written simply enough to introduce the HT area to "just plain folks" (non-specialist readers). We will try to make it readable for a person of first-degree level in

---

Hilbert Transform Applications in Mechanical Vibration, First Edition. Michael Feldman.
© 2011 John Wiley & Sons, Ltd. Published 2011 by John Wiley & Sons, Ltd.

Engineering Science who can understand the concept of the HT sufficiently to utilize it, or at least to determine if he or she needs to dig more deeply into the subject.

The book is divided into three main parts. The first describes the HT, the analytic signal, and the main notations, such as the envelope, the instantaneous phase, and the instantaneous frequency, as well as the analytic signal representation in the complex plane. This part also discusses the existing techniques for the HT realization in digital signal processing.

The second part describes the measured signal as a function of time, mostly vibration, which carries some important information. The HT is able to extract this time-varying information for narrow- and wideband signals. It is also capable of decomposing a multicomponent nonstationary signal into simple components or, for example, separate standing and traveling waves.

The third part is concerned with a mechanical system as a physical structure that usually takes an impulse or another input force signal and produces a vibration output signal. Use of the HT permits us to estimate the linear and nonlinear elastic and damping characteristics as instantaneous modal parameters under free and forced vibration regimes.

The book is a guide to enable you to do something with the HT, even if you are not an expert specializing in the field of modern vibration analysis or advanced signal processing. It should help you significantly (a) to reduce your literature research time, (b) to analyze vibration signals and dynamic systems more accurately, and (c) to build an effective test for monitoring, diagnosing, and identifying real constructions.

1.1   Brief history of the Hilbert transform

To place the HT subject in a historical context of mechanical vibration we will start with a very short chronology of the history of the HT. A traditional classical approach to the investigation of signals can include a spectral analysis based on the Fourier transform and also a statistical analysis based on a distribution of probabilities and other representations typical for random data. In addition to these typical spectral, correlation, and distribution characteristics, another method of representing and investigation a signal originated in the forties of the previous century (Gabor, 1946). –. This new method suggested the use of a random signal x as a product of two other independent functions: $x = A \cos \varphi$, where A is the amplitude, or envelope, and φ is the instantaneous phase. Thus, the variable x can be presented in the form of a harmonic fluctuation modulated in the amplitude and in the phase. This means of representing a function has appeared to be more descriptive and convenient for the solution of a number of theoretical and practical problems.

At that time researchers and engineers were not familiar with the HT (Therrien, 2002). However, they started to investigate the envelope and instantaneous phase by describing the signal in a $x - y$ Cartesian coordinate system (Bunimovich, 1951). In this $xy-$ plane the original signal was a first (x-axis) projection of the vector with length A and phase angle φ. The second projection in the $xy-$ plane along a vertical axis took the form $y = A \sin \varphi$. Due to the orthogonality of the bases, one obtains the following relations: $A^2 = x^2 + y^2$, $\varphi = \arctan\left(y/x\right)$. The same relations were extended to the case of a representation of a variable in the form of a Fourier series: $x = \sum\left(a_k \cos\varphi_k + b_k \sin\varphi_k\right)$, where each component of the sum means a simple harmonic. The mathematical literature (Titchmarsh, 1948) defined the second projection of the vector sum as the conjugate Fourier series $y = \sum\left(a_k \sin\varphi_k - b_k \cos\varphi_k\right)$. This started a study of the modulated signal, its envelope, instantaneous phase, and frequency based on the well-known Euler's formula for harmonic functions, according to which $e^{i\varphi} = \cos\varphi + i\sin\varphi$.

Nevertheless, a question of how an arbitrary (but not harmonic) signal should be represented to define the envelope and other instantaneous characteristics was still open. This problem was solved by Denis Gabor in 1946 when he was the first to introduce the HT to a signal theory (Gabor, 1946). Gabor defined a generalization of the Euler formula $e^{i\varphi} = \cos(\varphi) + i\sin(\varphi)$ in a form of the complex function $Y(t) = u(t) + iv(t)$, where $v(t)$ is the HT of $u(t)$. In signal processing, when the independent variable is time, this associated complex function is known as an analytic signal and the projection $v(t)$ is called a quadrature (or a conjugate) of the original function $u(t)$. The HT application to the initial signal provides some additional important information on an amplitude, instantaneous phase, and frequency of vibrations.

The analytic signal theory was then progressively developed by experts in various fields, mainly in electronics, radio, and physics. Here we must mention an important result called a Bedrosian condition (identity, equality), derived in 1963 (Bedrosian, 1963). This simplifies the HT calculation of a product of functions, helps us to understand the instantaneous amplitude and frequency of signals, and provides a method of constructing basic signals in the time–frequency analysis.

The theory and the HT application progressed greatly during the following years owing to Vakman, who further developed the analytic signal theory by solving problems of nonlinear oscillation and wave separation (Vakman and Vainshtein, 1977; Vainshtein and Vakman, 1983).

Investigators of digital algorithms of the HT realization (Thrane *et al.*, 1984) made a major contribution when a "digital" revolution started, and digital computers and digital signal procedures appeared everywhere. In 1985 Bendat suggested the inclusion of the HT as a typical signal procedure to the Brüel and Kjær two-channel digital analyzer. He also wrote a B&K monograph with a cover picture of David Hilbert's face gradually rotated through 90° (Bendat, 1985). As the speed and volume of digital processors keep increasing, software and digital hardware are replacing traditional analog tools, making today's devices smarter, more reliable, less expensive, and more power efficient than ever before.

The HT and its properties have been studied extensively in fluid mechanics and geophysics for ocean and other wave analysis (Hutchinson and Wu, 1996). A detailed analysis of the HT and complex signals was made by Hahn in 1996 (Hahn, 1996a). His book covers the basic theory and practical applications of HT signal analysis and simulation in communication systems and other fields. Two volumes of the *Hilbert Transforms* recently published by King (2009) are a very definitive reference on the HTs, covering mathematical techniques for evaluating them, and their application.

In 1998 an outstanding work by Huang gave a new push to the modern research in the field of HTs (Huang *et al.*, 1998). His original technique, known as the Empirical

Mode Decomposition (EMD), adaptively decomposes a signal into its simplest intrinsic oscillatory modes (components) at the first stage. Then, at the second stage, each decomposed component forms a corresponding instantaneous amplitude and frequency. Signal decomposition is a powerful approach; it has become extremely popular in various areas, including nonlinear and nonstationary mechanics and acoustics.

1.2   Hilbert transform in vibration analysis

In the field of radio physics and signal processing, the HT has been used for a long time as a standard procedure. The HT and its properties – as applied to the analysis of linear and nonlinear vibrations – are theoretically discussed in Vakman (1998). The HT application to the initial signal provides some additional information about the amplitude, instantaneous phase, and frequency of vibrations. The information was valid when applied to the analysis of vibration motions (Davies and Hammond, 1987). Furthermore, it became clear that the HT also could be employed for solving an inverse problem – the problem of vibration system identification (Hammond and Braun, 1986).

The first attempts to use the HT for vibration system identification were made in the frequency domain (Simon and Tomlinson, 1984; Tomlinson, 1987). The HT of the Frequency Response Function (FRF) of a linear structure reproduces the original FRF, and any departure from this (e.g., a distortion) can be attributed to nonlinear effects. It is possible to distinguish common types of nonlinearity in mechanical structures from an FRF distortion.

Other approaches (Feldman, 1985) were devoted to the HT application in the time domain, where the simplest natural vibration system, having a mass and a linear stiffness element, initiates a pure harmonic motion. A real vibration always gradually decreases in amplitude owing to energy losses from the system. If the system has nonlinear elastic forces, the natural frequency will depend decisively on the vibration amplitude. Energy dissipation lowers the instantaneous amplitude according to a nonlinear dissipative function. As nonlinear dissipative and elastic forces have totally different effects on free vibrations, the HT identification methodology enables us to determine some aspects of the behavior of these forces. For this identification in the time domain, it was suggested that relationships should be formed between the damping coefficient (or decrement) as a function of amplitude and between the instantaneous frequency and the amplitude. Lately, it was suggested that the linear damping coefficient could be formed by extracting the slope of the vibration envelope (Hammond and Braun, 1986; Agneni and Balis-Crema, 1989).

Some studies (Feldman, 1994a, 1994b), provide the reader with a comprehensive concept for dealing with a free and forced response data involving the HT identification of SDOF nonlinear systems under free or forced vibration conditions. These methods, being strictly nonparametric, were recommended for the identification of an instantaneous modal parameter, and for the determination of a system backbone and damping curve.

A recent development of the HT-based methods for analyzing and identifying single- and multi-DOF systems, with linear and nonlinear characteristics, is attributable to J.K Hammond, G.R. Tomlinson, K. Worden, A.F. Vakakis, G. Kerschen, F. Paia, A. R. Messina, and others who explored this subject much further. Since the HT application in the vibration analysis was reported only 25 years ago, it has not been well perceived in spite of its advantages in some practical applications. At present, there is still a lot to be done for both a theoretical development and practical computations to provide many of various practical requirements.

1.3   Organization of the book

This book proceeds with three parts and twelve chapters.

Part I "Hilbert Transform and Analytic Signal," contains three chapters. Chapter 1 gives a general introduction and key definition, and mentions concisely some of the HT history together with its key properties. Chapter 2, which includes a review of some relevant background mathematics, focuses on a rigorous derivation of the HT envelope and the instantaneous frequency, including the problem of their possible negative values. Chapter 3 deals with two demodulation techniques: the envelope and instantaneous frequency extraction, and the synchronous signal detection. It describes a realization of the Hilbert transformers in the frequency and time domains. The sources and characteristics of possible distortions, errors, and end-effects are discussed.

Part II, "Hilbert Transform and Vibration Signals," contains four chapters. Chapter 4 introduces typical examples of vibration signals such as random, sweeping, modulated, and composed vibration. It explains the derivatives, the integral, and the frequency content of the signal. Chapter 5 covers some new ideas related to the mono- and multicomponent vibration signal. Material that has important practical applications in signal analysis is treated, and some topics – especially the congruent envelope of the envelope – that have the potential for important practical applications are covered. Chapter 6 is devoted to examining the behavior of local extrema and the envelope function. Material in this chapter has an application to the explanation of the well-known Empirical Mode Decomposition (EMD). It also describes a relatively new technique called the Hilbert Vibration Decomposition (HVD) for the separation of nonstationary vibration into simple components. The chapter illustrates some limitations of the technique including the poor frequency resolution of the EMD. Most of key properties of these two decomposition methods are covered, and the most important application for typical signals is treated. Chapter 7 provides examples of HT applications to structural health monitoring, the real-time kinematic separation of nonstationary traveling and standing waves, the estimation of echo signals, a description of phase synchronization, and the analysis of motion trajectory.

Part III, "Hilbert Transform and Vibration Systems," contains five chapters. Chapter 8 gives some introductory material on quadrature methods, when the real and imaginary parts of a complex frequency function are integrally linked together by the HT. The chapter explains the important Kramers–Kronig formulas, used widely in applications. It covers some solutions of the frequency response function that can be used for the detection of nonlinearity. This chapter links both the initial nonlinear spring and the initial nonlinear friction elements and analytic vibration behavior. Both simple and mathematically rigorous derivations are presented. The chapter also covers some typical nonlinear stiffness and damping examples. Chapter 9 describes the foundation for the identification methods that are treated in the next chapter. The important sum rules that come directly from the HT relations – such as skeleton and damping curves, static force characteristics, and nonlinear output frequency response functions – are discussed in detail. Chapter 10 presents FREEVIB and FORCEVIB methods as a summary of all the key properties of the HT for practical implementation in dynamic testing. The skeleton and damping curves are treated together with the reconstructed initial nonlinear static forces. Chapter 11 treats the case of precise nonlinear vibration identification. The special difficulties that arise for the significant role of the large number of high-order superharmonics are analyzed in detail. Applications of some results developed in Chapter 9 for the identification of multi-degree-of-freedom (MDOF) systems are illustrated. Chapter 12, the final chapter, considers industrial applications in a number of different areas. To conclude the book, this chapter provides references to HT examples of a successful realization of the parametric and nonparametric identification of nonlinear mechanical vibration systems.

Part I

HILBERT TRANSFORM AND ANALYTIC SIGNAL

2

Analytic signal representation

2.1 Local versus global estimations

A measured varying signal can be described by different signal attributes that change over time. Estimating these attributes of a signal is a standard signal processing procedure. The HT provides the signal analysis with some additional information about amplitude, instantaneous phase, and frequency. To estimate the attributes – such as amplitude or frequency, – any procedure will need some measurements during a de nite time. Two approaches exist for such an estimation: a local approach that measures attributes at each instant without knowing the entire function of the process; and a global approach that depends on the whole signal waveform during a long (theoretically in nite) measuring time (Vakman and Vainshtein, 1977). An example of the local (or differential, microscopic) approach is a function extreme value estimation. A further example of the local approach is frequency estimation by measuring the interval (spacing) between two successive zero crossings.

The global (or integral, macroscopic) approach is something different. The following are examples of the global approach: estimating an average frequency by taking the rst moment of the spectral density, or estimating the mean value or a standard deviation of a function. In other words, local estimations consider the signal locally, that is, in a very small interval around the instant of the analysis. Quite to the contrary, global estimations have to use the whole measured signal. The HT as the subject of our examination is a typical example of the global approach. The global versus local estimations provide different precisions and resolutions depending on many conditions – primarily, noise distortions and a random icker phase modulation in a signal (Girolami and Vakman, 2002; Vakman, 2000).

2.2 The Hilbert transform notation

The HT is one of the integral transforms (like Laplace and Fourier); it is named after David Hilbert, who rst introduced it to solve a special case of integral equations

Hilbert Transform Applications in Mechanical Vibration, First Edition. Michael Feldman.
© 2011 John Wiley & Sons, Ltd. Published 2011 by John Wiley & Sons, Ltd.

in the area of mathematical physics (Korpel, 1982). The HT of the function $x(t)$ is de ned by an integral transform (Hahn, 1996a):

$$H\left[x(t)\right] = \tilde{x}(t) = \pi^{-1} \int_{-\infty}^{\infty} \frac{x(\tau)}{t-\tau} d\tau \qquad (2.1)$$

Because of a possible singularity at $t = \tau$, the integral has to be considered as a Cauchy principal value. The HT of a real-valued function $x(t)$ extending from $-\infty$ to $+\infty$ is a real-valued function $\tilde{x}(t)$ de ned by (2.1).

The mathematical integral de nition there really does not give much insight into the understanding and application of the HT. However, the physical meaning of the HT helps us to gain a much deeper access to the transformation. Physically, the HT is equivalent to a special kind of linear lter, where all the amplitudes of the spectral components are left unchanged, but their phases are shifted by $\pi/2$ (Figure 2.1c) (Hahn, 1996a; King, 2009; Thomas and Sekhar, 2005). Thus, the HT representation $\tilde{x}(t)$ of the original function is the convolution integral of $x(t)$ with $(\pi t)^{-1}$, written as $\tilde{x}(t) = x(t) * (\pi t)^{-1}$. The impulse response function of the ideal HT is shown in Figure 2.1a; the module and the phase characteristics of the HT transfer function are shown in Figure 2.1b and c.

It is clear that the HT of a time-domain signal $x(t)$ is another time-domain signal $\tilde{x}(t)$, and if $x(t)$ is real valued, then $\tilde{x}(t)$ is also real valued.

2.3   Main properties of the Hilbert transform

The HT is a linear operator, so if a_1 and a_2 are arbitrary (complex) scalars, and $x_1(t)$ and $x_2(t)$ are varying signals, then $H\left[a_1 x_1(t) + a_2 x_2(t)\right] = a_1 \tilde{x}_1(t) + a_2 \tilde{x}_2(t)$.

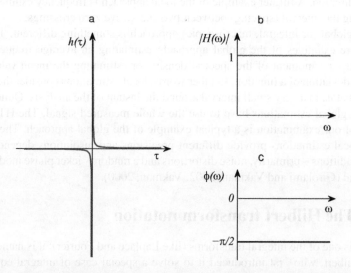

Figure 2.1   The ideal HT: the impulse response function (a), the module (b), and the phase (c) of the HT transfer function (Feldman, ©2011 by Elsevier)

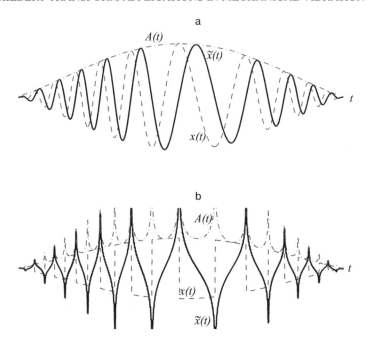

Figure 2.2   The quasiharmonic (a) and the square wave (b): the initial signal $x(t)$, the HT pair projection $\tilde{x}(t)$, the envelope $A(t)$ (Feldman, ©2009 by John Wiley & Sons, Ltd.)

In particular, the HT of a constant is zero. The double HT (the HT of a HT) yields the original function with the opposite sign, hence it carries out a shifting of the initial signal by $-\pi$. The HT used four times on the same real function returns the original function. The power (or energy) of a signal and its HT are equal. A function and its HT are orthogonal over an in nite interval $\int_{-\infty}^{\infty} x(t)\tilde{x}(t)dt = 0$. The HT of a function derivative is equivalent to the derivative of the HT of the function. The HT of a sine function is a cosine function; the HT of a cosine function is a negative sine function, but for some dissimilar waveform it can have a more complicated form (Figure 2.2).

A signal $x(t)$ and its HT projection $\tilde{x}(t)$ have the same amplitude spectrum and the same autocorrelation function. The reader with applications of signal processing in mind can nd further, more detailed, information on the properties of the HT – with a thorough explanation and a stringent presentation of their physical meaning in Johansson (1999) and King (2009).

2.4   The Hilbert transform of multiplication

In practice, a need for the HT of a functions product arises quite often. The equation

$$H[n_{slow}(t)x_{fast}(t)] = n_{slow}(t)\tilde{x}_{fast}(t),\qquad(2.2)$$

called a Bedrosian identity (product theorem), simpli es calculations of the HT of a function product (Bedrosian, 1963). This states that the HT of the product of lowpass and highpass signals with nonoverlapping spectra is de ned by a product of the lowpass signal and the HT of the highpass signal.

For a more complicated case of a product of real functions with overlapping spectra we can use another formula (Hahn, 1996b). This formula for the HT of the product of overlapping spectra is derived by decomposing the signal into a sum of two parts with lowpass and highpass terms, thus enabling an application of Bedrosian theorem to each of the parts separately.

Let $n(t)$ and $x(t)$ be fast-varying functions whose frequency bands overlap. If one of the functions can be represented in the form of a sum of two parts $n(t) = \bar{n}_{slow}(t) + /_{fast}(t)$, then the HT of the product of these functions with overlapping spectra can also be written in the form of a sum of two parts (Hahn, 1996b):

$$H\left[n(t)x(t)\right] = H\left\{\left[\bar{n}_{slow}(t) + \ddot{n}_{fast}(t)\right]x(t)\right\} = \bar{n}_{slow}(t)\tilde{x}(t) + \tilde{n}_{fast}(t)x(t), \qquad (2.3)$$

where $\bar{n}_{slow}(t)$ is the slow (lowpass) part of the real function, $\ddot{n}_{fast}(t)$ is the fast (highpass) part, and $\tilde{n}_{fast}(t)$ is the HT pair component of the fast component $\ddot{n}_{fast}(t)$. This means that the HT of the signal multiplication with overlapping spectra results in a composition of two multiplications.

For example, the HT of the square of the harmonic $x^2 = (\cos\varphi)^2$ is equal to $H[x^2] = H[xx] = H\left[(0 + x)x\right] = 0 + \tilde{x}x = \sin\varphi\cos\varphi = \frac{1}{2}\sin2\varphi$. Another example for the cube of the harmonic $x^3 = (\cos\varphi)^3$ gives $H\left[x^3\right] = H\frac{1}{2}\tilde{x} + H\left[x^2\right]x = \frac{1}{2}\sin\varphi + \frac{1}{2}\sin2\varphi\cos\varphi = = \frac{1}{4}(3\sin\varphi + \sin3\varphi)$. These simple examples can be veri ed using complex numbers. It is known that for real numbers $x\,e^{ix} = \cos x + i\sin x$, so $e^{-ix} = \cos(-x) + i\sin(-x) = \cos x - i\sin x\cos x - i\sin x$. Adding these two equations yields $\cos x = \frac{e^{ix}+e^{-ix}}{2i}$. For the cube example we will have $\cos^3 x = \left(\frac{e^{ix}+e^{-ix}}{2i}\right)^3 = -\frac{e^{3ix}+e^{-3ix}+3(e^{ix}+e^{-ix})}{8i}$, or $\cos^3 x = \frac{1}{4}(3\cos x + \cos 3x)$. Therefore the HT of the cube will be $H\left[\cos^3 x\right] = \frac{1}{4}H\left[3\cos x + \cos 3x\right] = \frac{1}{4}(3\sin x + \sin 3x)$ that is, equal to the result obtained with the generalized HT of the product of the overlapping functions (2.3).

2.5   Analytic signal representation

The complex signal whose imaginary part is the HT (2.1) of the real part is called an *analytic* or *quadrature* signal (Lyons, 2000; Vakman, 1998). It is a two-dimensional signal whose value at some instant in time is speci ed by two parts, a real part and an imaginary part (Schreier and Scharf, 2010):

$$X(t) = x(t) + i\tilde{x}(t), \qquad (2.4)$$

where $\tilde{x}(t)$ is related to $x(t)$ by the HT. An example of a complex trace of the analytic signal uniquely de ned by Equation (2.1) is shown in Figure 2.3 as a helix that spirals around the time axis (Lyons, 2000).

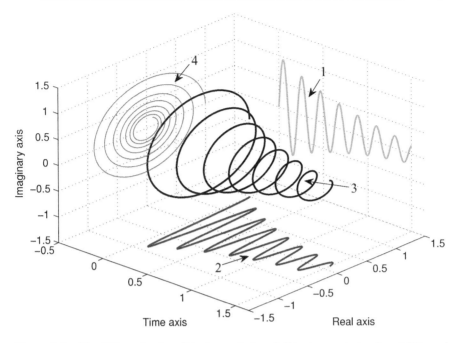

Figure 2.3   The HT projection (1), the real signal (2), the analytic signal (3), and the phasor in complex plain (4) (Feldman, ©2011 by Elsevier)

For each integral transform, there will be another relation which converts the transformed function back into the original function. At the same time this relation is usually an integral transform, although sometimes it may be written in terms of algebraic operations only. In order to return from a complex form of the analytic signal $X(t)$ back to the real function $x(t)$, one has to use a substitution $x(t) = 0.5 \, [X(t) + X^*(t)]$, where $X^*(t)$ is the complex conjugate signal of $X(t)$ (Vainshtein and Vakman, 1983). The analytic signal has a one-sided spectrum of positive frequencies. The conjugate analytic signal has a one-sided spectrum of negative frequencies.

2.6   Polar notation

According to analytic signal theory, a real vibration process $x(t)$, measured by, say, a transducer, is only one of many possible projections (the real part) of some analytic signal $X(t)$. The second, or quadrature, projection of the same signal (the imaginary part) $\tilde{x}(t)$ will then be conjugated according to the HT (2.1). An analytic signal has a geometrical representation in the form of a phasor rotating in the complex plane, as shown in Figure 2.4.

A *phasor* can be viewed as a vector at the origin of the complex plane having a length $A(t)$ and an angle, or an angular position (displacement), $\psi(t)$. The projection on the real axis is the initial real signal and is described by $x(t) = A(t)\cos\psi(t)$. Using

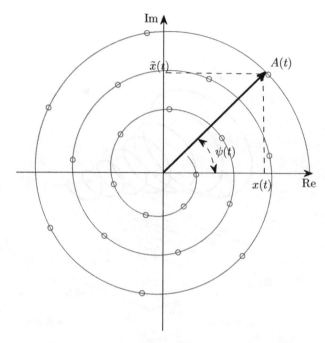

Figure 2.4 The analytic signal in the complex plain (Feldman, ©2009 by John Wiley & Sons, Ltd.)

a traditional representation of the analytic signal in its trigonometric or exponential form (Vainshtein and Vakman, 1983; Hartmann, 1998)

$$X(t) = |X(t)| [\cos \psi(t) + j \sin \psi(t)] = A(t)e^{j\psi(t)} \qquad (2.5)$$

one can determine its instantaneous amplitude (envelope, magnitude, modulus)

$$A(t) = \pm |X(t)| = \pm\sqrt{x^2(t) + \tilde{x}^2(t)} = \pm e^{\mathrm{Re}[\ln X(t)]} \qquad (2.6)$$

and its instantaneous phase

$$\psi(t) = \arctan \frac{\tilde{x}(t)}{x(t)} = \mathrm{Im}[\ln X(t)]. \qquad (2.7)$$

The change of coordinates from rectangular $(x, \tilde{x})$ to polar (A, ψ) produces $x(t) = A(t)\cos\psi(t)$, $\tilde{x}(t) = A(t)\sin\psi(t)$.

2.7 Angular position and speed

An *angular position* of the phasor ψ is its angle relative to a xed direction, which we take as the zero angular position. A pure rotation produces no change in the vector length relative to the origin. The velocity of the rotating phasor has a cross-radial component perpendicular to the radius-vector that is called angular speed. The *angular speed* ω (also referred to as rotational speed, angular frequency, circular frequency, orbital frequency, transverse velocity, or radian frequency) is a rotational velocity and its magnitude is a scalar measure of the rotation rate. During a time interval $\Delta t = t_2 - t_1$ an average angular speed $\bar{\omega}$ is de ned as a ratio of an angular displacement to the time in which it occurs: $\bar{\omega} = \Delta\psi/\Delta t = (\psi_2 - \psi_1)/(t_2 - t_1)$. In the same way that linear velocity is the rst derivative of linear displacement, angular velocity is the rst derivative of angular position $\omega = \psi$. The modulus of the rst derivative $|\omega|$ measures how fast an object is rotating. The sign of the angular frequency (the sign of ω) indicates the rotation direction; for example, when $\omega > 0$, the vector rotates counterclockwise, and when $\omega < 0$, the vector rotates clockwise with time.

Due to the envelope variation in time, the phasor velocity has also a radial component along the radius de ned as the component of velocity away from or toward the origin. Geometrically, ψ is given by $\psi = s/A$, where s is the length of the circular arc that extends from the x-axis (the zero angular position) to the reference vector, and A is the vector length. The angular velocity of a phasor can be related to its full *translational velocity* ds/dt, which is a vector quantity and depends on both the length of the phasor A (the distance from the center of rotation) and the angular velocity ω. For the arc length $s = A\psi$ and the radius vector A we can write: $s = (A\psi)' = A(t)\omega(t) + \psi(t)A(t)$.

2.8   Signal waveform and envelope

Geometrically, the envelope means an integral curve which determines a singular position of the initial function (Kultyshev, 1990). The envelope may be constant (in which case the wave is a continuous harmonic) or may vary with time. The form or the shape of the variation of the instantaneous amplitude is called a wave envelope. If the waveform is a pure harmonic with varying positive and negative signal values, the relationships between peak-to-peak amplitude and the root mean square (the standard deviation) values are xed and known, as they are for any continuous periodic wave: Peak-to-peak = $2\sqrt{2}$RMS, where RMS = $[\Delta t^{-1} \int_0^{\Delta t} x^2(t)dt]^{\frac{1}{2}}$. However, this is not true for an arbitrary waveform which may or may not be periodic or continuous.

In a general case an amplitude of the oscillation $A(t)$ – as a magnitude of the complex analytical signal – varies with time (2.6). An initial signal and its envelope have common tangents at points of contact, but the signal never crosses the envelope.

Usually the envelope is considered positive, but that is not the issue. In the same way that the square root of a number can be negative, the real-valued envelope can have negative values. The plus sign of the root square corresponds to the *upper positive envelope* $+A(t)$, and the minus sign corresponds to the *lower negative envelope* $-A(t)$, so they are always in an antiphase relation.

An envelope function contains important information about the energy of the signal. It is known that power is a time average of energy (energy per unit time):

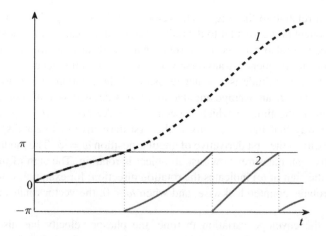

Figure 2.5 The instantaneous phase: unwrapped (- -) and wrapped (—) (Feldman, ©2009 by John Wiley & Sons, Ltd.)

Power $= T^{-1} \int_0^T x^2(t) = \text{RMS}^2$. Because the power in the HT pair projection is equal to the power in the original signal, the envelope power is precisely twice the signal power.

By using the HT, the rapid oscillations can be removed from the amplitude modulated signal to produce a direct representation of the slow envelope alone. In some cases of over-amplitude modulation the signal modulated amplitude can be presented as an oscillated complex envelope function having negative values (see Section 4.5.3).

2.9 Instantaneous phase

For any signal there is a unique single value of its instantaneous phase at any given time that de nes where vector is pointing. The instantaneous orientation phasor angle de ned by Equation (2.7) is measured in radians (rad) rather than revolutions (rev), or degrees. This instantaneous phase notation, based on the arctangent, indicates a multibranch character of the function, as shown in Figure 2.5 (solid line), when the phase angle jumps between π and $-\pi$. We do not reset ψ to zero with each complete rotation of the reference line about the rotation axis. If the reference line completes two revolutions from the zero angular position, then the angular position ψ of the line is $\psi = 4\pi$ rad. These phase jumps can be unwrapped into a monotone function by changing the phase values arti cially (see Figure 2.5, dashed line).

An instantaneous relative phase shift in the case of two different but narrowband signals $x_1(t)$ and $x_2(t)$ can be estimated as an instantaneous relative phase between them, according to the formula (Feldman, 2001):

$$\Delta\psi = \psi_2(t) - \psi_1(t) = \arctan\frac{x_1(t)\tilde{x}_2(t) - \tilde{x}_1(t)x_2(t)}{x_1(t)x_2(t) + \tilde{x}_1(t)\tilde{x}_2(t)}. \tag{2.8}$$

The relative phase shift $\Delta\psi = \psi_2 - \psi_1$ is a function of the signal frequency $\Delta\psi(\omega)$; the relative phase shift for the specified frequency is associated with the *time delay* between two signals $\Delta t = \Delta\psi(\omega)/\omega$. There exists a *group delay* defined as a negative derivative of the phase angle $\Delta t_{group} = -d\,[\Delta\psi(\omega)]/d\omega$ for any narrowband signal with a group of harmonics around the frequency ω (Shin and Hammond, 2008; Perry and Brazil, 1997). In the case of two arbitrary signals the time delay between them can be estimated by the HT algorithms (Hertz, 1986).

2.10   Instantaneous frequency

More than several decades ago, Gabor (1946) and Ville (see Boashash, 1992) defined the notions of the phase and instantaneous frequency (IF) of a signal via an analytic signal. The first derivative of the instantaneous phase as a function of time

$$\omega(t) = \dot{\psi}(t) \tag{2.9}$$

– called the instantaneous angular frequency – plays an important role in signal analysis. For any signal there is a unique single value of the instantaneous phase at any given time. The dimension of the angular frequency $\omega(t) = 2\pi f(t)$ is in radians per second, and the cycle frequency $f(t)$ is in Hertz. There is a simple way to avoid the whole phase-unwrapping problem if you find the IF by differentiation of the signal itself

$$\omega(t) = \frac{x(t)\dot{\tilde{x}}(t) - \dot{x}(t)\tilde{x}(t)}{A^2(t)} = \mathrm{Im}\left[\frac{\dot{X}(t)}{X(t)}\right] \tag{2.10}$$

The IF $\omega(t)$ measures the rate and direction of a phasor rotation in the complex plane. Naturally, for a simple monoharmonic signal, the envelope and the IF are constant, and the phase angle increases linearly with time. In a general case, the IF of the signal is a varying function of time. Moreover, the IF in some cases may change sign in some time intervals, which corresponds to a change in the phasor rotation from counterclockwise to clockwise. The IF always has a simple and clear physical meaning – it is no more than just a varying speed (rate) of the phasor rotation in polar axes. In the time domain the negative IF corresponds to the appearance of a complicated riding cycle of an alternating signal.

At each moment a signal has only one single value of the IF (Loughlin and Tacer, 1997). For nonstationary signals (that is, signals whose spectral contents vary with time) the IF plays an important role and can be estimated by different algorithms (Vakman, 2000). These algorithms of the IF direct nonparametric estimation are based on definitions (2.9) and (2.10), and do not require any other time–frequency analysis or any *a priori* mathematical model of the signal.

Algorithm 1, based on formula (2.9), means differentiation of the phase angle. However, an initial arctangent function always produces sharp jumps between π and $-\pi$ values and cannot be used for a direct differentiation. Therefore, prior to the differentiation, the algorithm performs a phase unwrapping procedure to attain a continuous phase. The unwrapping procedure produces a smooth, increasing phase function by adding 2π every time a full cycle is completed. Immediately after the phase unwrapping, the algorithm performs its differentiation according to (2.9).

Algorithm 2, presented by formula (2.10), is obtained by an analytical differentiation of the arctangent of the fraction $\omega(t) = d\left[\arctan \frac{\tilde{x}(t)}{x(t)}\right]/dt$. Thus, it does not operate with the phase function but, instead uses the initial signal $x(t)$, its HT projection $\tilde{x}(t)$, and their rst derivatives $x(t)$, $\tilde{x}(t)$.

Algorithm 3, the simplest procedure, takes into account the discrete form of a real signal $x(n)$ obtained by sampling the analog signal at discrete instants of time t_n. The corresponding analytic signal will have a form $X_n = x_n + i\tilde{x}_n$. The algorithm also uses a conjugate complex signal at the next discrete instants of time t_{n+1}: $X_{n+1}^* = x_{n+1} - i\tilde{x}_{n+1}$. As the rst difference of the phase, the IF can be calculated as a symbolic difference of two arctangents $\psi = \Delta\psi_n/\Delta t$ between two adjacent samples of the phase angle ($\Delta t = 1$):

$$\Delta\psi_n = \psi_{n+1} - \psi_n = \arctan(\tilde{x}_{n+1}/x_{n+1}) - \arctan(\tilde{x}_n/x_n)$$
$$= \arctan\frac{\tilde{x}_{n+1}x_{n+1} - \tilde{x}_n/x_n}{1 + \tilde{x}_{n+1}\tilde{x}_n/x_{n+1}x_n} = \arctan\frac{\tilde{x}_nx_{n+1} - x_n\tilde{x}_{n+1}}{x_nx_{n+1} + \tilde{x}_n\tilde{x}_{n+1}}$$

The multiplication of the initial analytic signal and the conjugate complex signal produces a new complex function $X_nX_{n+1}^* = (x_n - i\tilde{x}_n)(x_{n+1} - i\tilde{x}_{n+1}) = x_nx_{n+1} + \tilde{x}_n\tilde{x}_{n+1} + i(\tilde{x}_nx_{n+1} - x_n\tilde{x}_{n+1})$ whose angle is equal to the IF of the signal. For $\Delta t = 1$ we can therefore write:

$$\psi = \Delta\psi_n = \arctan\left(X_nX_{n+1}^*\right) \tag{2.11}$$

The above equation allows us to compute the IF directly by computing the arctangent of the conjugate multiplication of adjacent complex samples. Unwrapping is not required with this algorithm. Direct estimation of the IF by using the simple formula (2.11) is a nice alternative to differentiation of the signal instantaneous phase.

Other, more complicated, algorithms for the IF estimation in the case of frequency modulated signals are discussed in, for example, Goswami and Hoefel (2004). These algorithms are based on the HT, Haar wavelet, and generalized pencil of function methods. While Algorithm 3 appears to be least sensitive to noise, the method described in Algorithm 1 is the easiest to implement. The wavelet-based method is also computationally more ef cient and can be implemented in real-time.

2.11   Envelope versus instantaneous frequency plot

Each of the instantaneous functions $A(t)$, $\omega(t)$ is a parametric function of time (with the parameter t) which is regularly plotted on a separate graph. This way we achieve the best possible time and frequency resolution of analysis. The instantaneous amplitude and the IF as functions of time can also be represented in a 3D plot.

Combining the envelope and the IF time functions, and excluding the parameter, we will get a 2D plot determined by a set of pairs of A, ω. Instantaneous functions plotted in conjunction can be shown with linear or logarithm axes. The envelope versus the IF plot shows the instantaneous characteristics relationship and is especially useful for identifying a vibration system (Feldman, 1985). For example, constant terms such as $\omega =$ constant contribute a straight vertical line of the unvarying resonance frequency of the linear vibration system. Some other examples of typical envelope–frequency plots are discussed in Chapter 4.

2.12   Distribution functions of the instantaneous characteristics

The envelope and the IF of real vibration signals are nonconstant; they can vary randomly in time. Unlike deterministic signals, the behavior of random signals can be analyzed with probability and statistical functions. Taking into account the analytic signal representation enables one to consider a vibration process, at any moment in time, as a quasiharmonic oscillation, that is amplitude and frequency modulated by time-varying functions $A(t)$ and $\omega(t)$:

$$x(t) = A(t)\cos \int_0^t \omega(t)dt. \tag{2.12}$$

The instantaneous parameters are functions of time and can be estimated at any point of the vibration signal. The total number of points that map the vibration is much larger than the number of peak points of the signal. It opens the way for averaging and for other statistical processing procedures, making a vibration analysis more precise.

2.12.1   Envelope distribution and average values

The distribution of an envelope appears as a distribution of random complex numbers whose real and imaginary Gaussian components are independent and identically distributed. In that case, the envelope as a modulus (an absolute value) of a complex number is Rayleigh-distributed (Whitaker, 2005). As an example of this relation, a typical classical Gaussian (normal) form of the probability density of the random vibration, which conforms to the Rayleigh probability density of the vibration envelope, is presented in Section 4.1 (Figure 4.1b).

A mean value (average) of the envelope takes the form $\overline{A} = T^{-1} \int_0^T A(t)dt = \mathrm{RMS}_x \sqrt{\pi/2} \approx \mathrm{RMS}_x \times 1.253$, where RMS is the root mean square (the standard deviation) of the random signal. The variance (the mean value of the square) of the envelope is

$$\overline{[A(t)]^2} = T^{-1} \int_0^T A^2(t)dt = (2 - \pi/2)\,\mathrm{RMS}_x^2 \approx 0.429 \times \mathrm{RMS}_x^2, \tag{2.13}$$

which determines the level of the envelope variation. Generally the envelope probability density $p(A)$ is related to the signal probability density function $p(x)$: $p(x) = \pi^{-1} \int_{|x|}^{\infty} \frac{p(A)dA}{(A^2-x^2)^{1/2}}$. Estimation examples of the envelope average for deterministic vibration are given in Section 5.4. Some more interesting composition possibilities – arising when summations are considered for harmonic, modulated, and random signals – are considered in Cain, Lever, and Yardim (1998).

2.12.2   Instantaneous frequency average values

The rst moments of the IF are closely related to the rst moments of the spectral density of a random signal (Boashash, 1992; Davidson and Loughlin, 2000). Let m_j be the j central spectral moment given by the complex spectrum $S(\omega)$: $m_j = (2\pi)^{-1} \int_0^{\infty} \omega^j |S(\omega)|^2 \, d\omega$. Thus the mean value of the IF, called a *central frequency* ω_0, will be equal to the rst normalized moment ($j = 1$) of the signal spectrum (Vainshtein and Vakman, 1983):

$$\omega_0 = \bar{\omega} = \int_{-\infty}^{\infty} \omega(t)A^2(t)dt = \frac{m_1}{m_0}. \tag{2.14}$$

The mean value of the IF squared $\overline{\omega^2} = \int_{-\infty}^{\infty} \omega^2(t)A^2(t)dt = \frac{m_2}{m_0} - \overline{A^2(t)}$ determines the level of the IF variation. Some typical examples of the central frequency estimation based on the IF are given in Section 4.1 (Table 4.1) and Section 5.4.

2.13   Signal bandwidth

There are several techniques for estimating the frequency bandwidth of a narrowband signal. Probably the most familiar, and simplest, is a half-peak level width of the signal spectrum. Also, a spectrum bandwidth could be estimated as the width of a hypothetical square with the same energy and peak value $\Delta\omega_1 = \int_0^{\infty} \omega \, S^2(\omega)d\omega/m_0$. In the case of IF analysis, it is useful to introduce a further width parameter that is equal to the mean absolute value of the IF deviation from its central value plus the envelope variations. By summing up the IF variations around the mean value and the envelope variations, we will obtain an average spectrum bandwidth of the signal $\Delta\omega_2$ (Fink, 1966; Cohen and Lee, 1989; Vainshtein and Vakman, 1983):

$$\Delta\omega_2^2 = (2\pi)^{-1} \int_0^{\infty} (\omega - \omega_0)^2 |S(\omega)|^2 \, d\omega = \overline{\omega^2} + \overline{A^2} - \bar{\omega}^2 = \frac{m_2}{m_0} - \left(\frac{m_1}{m_0}\right)^2 \tag{2.15}$$

where, again, m_j is the j th moment of the spectrum $S(\omega)$. Equation (2.15) is derived from Parseval's energy identity relation for time and frequency domains. It indicates that the signal spectrum bandwidth $\Delta\omega$ is equal to the sum of the IF $\overline{\omega^2}$ and the envelope $\overline{A^2}$ variations. For signals that are only amplitude modulated, the spectral bandwidth is equal to the mean square value of the rate of amplitude variation. For signals that are only frequency modulated, the spectral bandwidth is equal to the mean square value of the IF. Some typical examples of the spectral bandwidth estimation are presented in Table 4.1 and Sections 4.11 and 4.12.

2.14 Instantaneous frequency distribution and negative values

Since the IF may be considered to be a signal frequency at a given time instant by estimation of the length of the whole signal waveform, it seems reasonable to inquire about the IF probability density function, or spread, at that observation length. For the random normal narrowband signal, the probability density function of the IF was derived by Bunimovich (Bunimovich, 1951; Broman, 1981) $p(\omega) = \frac{\Delta\omega^2}{2(\omega^2 + \Delta\omega^2)^{3/2}}$. A probabilistic prerequisite to the formation of the negative value of the IF is then:

$$p\left[\omega(t) < 0\right] = 0.5\left(1 - \frac{\bar{\omega}}{|\omega|}\right). \tag{2.16}$$

For example, the probability of a negative value of the IF of a random signal after an ideal rectangular narrowband lter is directly proportional to the relative lter width $p\left[\omega(t) < 0\right] = \frac{\Delta\omega^2}{144\omega_0^2}$, where ω_0 is the central lter frequency, and $\Delta\omega$ is the lter width. This indicates that after any narrowband ltering the random signal will still have a negative value of IF. However, for very small widths like $\frac{\Delta\omega}{\omega_0} \leq 0.01$, the probability will be less than one in a million, and the IF can be almost always be considered positive.

2.15 Conclusions

The application of the HT to signal analysis provides some additional information about the amplitude, instantaneous phase, and frequency of vibration. This chapter discussed some basic mathematical properties of the HT and of the analytic signal, which allow one to determine the amplitude, phase, and frequency of any oscillation at any instant of time. For narrowband vibration and for slow frequency modulation, these instantaneous characteristics agree with the intuitive meaning of the signal amplitude, phase, and frequency. Considerable attention has been devoted to the meaning and conditions of existing negative frequency components in the IF. The negative IF is not "unphysical," it can be easily understood through a variation in phase angle. The IF always has a simple and clear physical meaning – it is no more than just a varying speed (rate) of the phasor rotation in polar axes. A negative IF corresponds to a change of phasor rotation from counterclockwise to clockwise. In the time domain the negative IF corresponds to the appearance of a complicated riding cycle of an alternating signal.

3

Signal demodulation

Traditional Fourier analysis simply assumes that the signal is a sum of a number of sine waves. The HT allows us to obtain a complex demodulation analysis, adapted to signals of the form of a single, but modulated (perturbed), sine wave (Claerbout, 1976). Since a vibration signal is exactly of that model, it is no wonder that in some cases the HT performs better than Fourier analysis. A demodulation removes the modulation from a signal and returns the original baseband signal. Thus demodulation is a process of extracting the original information incorporated in a modulated signal. For example, an envelope detector based on the HT takes a high-frequency signal as an input and provides an output that is an envelope of the original signal.

3.1 Envelope and instantaneous frequency extraction

An amplitude modulated wave $x(t) = A(t) \cos \psi(t)$ should be processed in some way to preserve only the modulating envelope function $A(t)$ and discard the oscillations $\cos \psi(t)$. The polar coordinates allow us to separate easily the effects of amplitude and phase (or frequency) modulation, and effectively demodulate certain kinds of signals. If we can generate a quadrature signal $\tilde{x}(t) = A(t) \sin \psi(t)$ of the same modulated signal, then we can easily generate the envelope from (2.6): $A^2(t) = A^2(t) \cos^2 \psi(t) + A^2(t) \sin^2 \psi(t)$. In this way, just by computing the square root of the sum of the squares of the real and imaginary parts, we obtain the envelope function at any time. The HT, as a $90°$ phase shift on every frequency component, is the best signal-processing procedure for obtaining the Hilbert component and estimating the instantaneous amplitude (Lyons, 2000, 2004). Such a HT envelope detector block diagram is shown in Figure 3.1. In fact, the envelope detectors can be as simple as only a diode and a lowpass lter, but the performance of the HT envelope detector is quite precise and it is not sensitive to the carrier. Moreover, the HT – in addition to the envelope – allows us to extract a carrier wave in the form of an instantaneous

Hilbert Transform Applications in Mechanical Vibration, First Edition. Michael Feldman.
© 2011 John Wiley & Sons, Ltd. Published 2011 by John Wiley & Sons, Ltd.

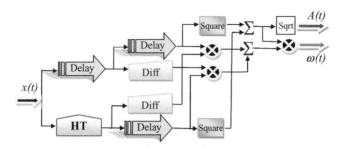

Figure 3.1   The block diagram of the envelope and the IF extraction (Feldman, ©2011 by Elsevier)

phase $\psi(t)$ or an IF $\omega(t) = \psi(t)$. A block diagram of a signal demodulation and an extraction of the IF on the base of formula (2.10) is also shown in Figure 3.1.

This demodulation block diagram converts an arbitrary unknown oscillation input $x(t)$ into two new functions: the envelope $A(t)$ and the IF $\omega(t)$. The diagram, which also includes the Hilbert transformer (HT), the differentiators (Diff), and the time delay blocks along with algebraic transformations, – can be used for a real-time signal demodulation.

The HT demodulation and extraction of the original instantaneous amplitude/frequency information from a modulated carrier wave operates with any oscillation signal. Mathematically, it is correct for any vibration, but in practice it is essential mostly for narrowband signals where it generates slow-varying instantaneous characteristics. Other types of input vibration, such as a wideband signal or a composition of several harmonics, will result in complicated fast-varying instantaneous characteristics. Such an output of the demodulation can become more complicated than the input signal itself, which does not make sense. In the case of a signal composition it is desirable to have an opportunity to demodulate each single speci c oscillating component even if it is a nonstationary oscillating function (see Sections 6.1 and 6.3).

3.2   Hilbert transform and synchronous detection

The signal synchronous demodulation technique is well known and has many names, including synchronous detection, in-phase/quadrature demodulation, coherent or heterodyne demodulation, auto-correlation, signal mixing and frequency shifting, lock-in ampli er detection, and phase sensitive detection (Whitaker, 2005). A synchronous demodulation considers the initial signal as a sum of components with a slow-varying instantaneous amplitude and frequency, so that $x(t) = \sum A_l(t) \cos \left(\int \omega_l(t)dt \right)$, where $A_l(t)$ is an instantaneous amplitude and $\omega_l(t)$ is the l-component IF. A synchronous demodulation – in addition to the initial composition $x(t)$ – considers that the frequency of the demodulated component is *given a priori* as a reference carrier frequency $\omega_r(t)$. In essence, a synchronous demodulation extracts the amplitude details about an oscillation component with a known frequency by multiplying the initial composition by two reference signals exactly $90°$ out of phase with one another. For the output we will get two projections, the in-phase and the HT (quadrature) phase output. The component amplitude can be obtained by taking the square root of sums of the squares of these projections.

In this case, a single oscillating component $x_{l=r}(t) = A_{l=r}(t) \cos \left(\int \omega_{l=r}(t) dt \right)$ on exactly the same frequency as the reference signal $\cos \left(\int \omega_r(t) dt \right)$ is mixed with other l components. The in-phase signal part $x_{l=r}(t)$ is given, as

$$
\begin{aligned}
x_{l=r}(t) &= \sum \left[A_l(t) \cos \left(\int \omega_l(t) dt + \varphi_l(t) \right) \right] * \cos \left(\int \omega_r(t) dt \right) \\
&= \frac{1}{2} A_l(t) \left[\cos \left(\int (\omega_l(t) - \omega_r(t)) \, dt + \varphi_l(t) \right) \right. \\
&\quad \left. + \cos \left(\int (\omega_l(t) + \omega_r(t)) \, dt + \varphi_l(t) \right) \right] \\
&= \frac{1}{2} A_l(t) \left[\cos (\varphi_l(t)) + \cos \left(\int (\omega_l(t) + \omega_r(t)) dt + \varphi_l(t) \right) \right], \quad (3.1)
\end{aligned}
$$

where $A_l(t)$, $\omega_l(t)$, and $\varphi_l(t)$ are the amplitude, the IF, and the phase angle of the l-component respectively, and $\omega_r(t)$ is the IF of the r-reference largest component.

The second phase-shifted quadrature part $\tilde{x}_{l=r}(t)$ is given by the analogous formula

$$
\tilde{x}_{l=r}(t) = \frac{1}{2} A_l(t) \left[-\sin (\varphi_l(t)) - \sin \left(\int (\omega_l(t) + \omega_r(t)) dt + \varphi_l(t) \right) \right].
$$

Each of the obtained parts consists of two different functions. One is a slow-varying function, which includes an amplitude and a phase, and the other is a fast-oscillating) part, which includes a double frequency harmonic. In such a case, it is again possible to remove the oscillating part by the use of lowpass ltering. Oscillating components that are not of the exact same frequency as the reference ($\omega_l \neq \omega_r$) will not yield this slow-varying function. Thus, only the slow part will be retained, and the amplitude and phase components can both be calculated:

$$
\langle x_{l=r}(t) \rangle = \begin{cases} \frac{1}{2} A_l(t) \cos \varphi_l(t), & \text{if } \omega_l = \omega_r \\ 0, & \text{if } \omega_l \neq \omega_r \end{cases}; \quad \langle \tilde{x}_{l=r}(t) \rangle = \begin{cases} -\frac{1}{2} A_l(t) \sin \varphi_l(t), & \text{if } \omega_l = \omega_r \\ 0, & \text{if } \omega_l \neq \omega_r \end{cases}
$$

$$
A_{l=r}(t) = 2\sqrt{\langle x_{l=r}(t) \rangle^2 + \langle \tilde{x}_{l=r}(t) \rangle^2}; \quad \varphi_{l=r}(t) = -\arctan \frac{\langle \tilde{x}_{l=r}(t) \rangle}{\langle x_{l=r}(t) \rangle}
$$

No matter what the instantaneous phase is, the resultant envelope $A_{l=r}(t)$ always represents the detected component envelope. A synchronous detection technique is capable of measuring even small varying signals that are obscured by large numbers of other components. A block diagram of the synchronous demodulation is shown in Figure 3.2. The demodulation block diagram extracts an unknown varying envelope $A(t)$ of the oscillation input $x(t)$ for the known reference IF $\omega(t)$. The diagram includes a Hilbert transformer (HT), time delay blocks, lowpass frequency lters (LPF), along with algebraic transformations and can be used to demodulate a real-time signal.

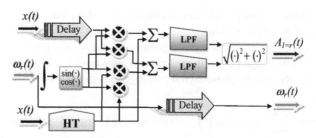

Figure 3.2   The block diagram of the synchronous demodulation (Feldman, ©2011 by Elsevier)

Synchronous detectors are not considerably more complex than simple HT detectors. A demodulation is performed by multiplying the modulated carrier by a wave, thus synchronous detectors are a subset of "product" detectors. The advantage of a synchronous detection is that it causes less distortion than an envelope detection and works well with single sideband signals. It is a preferred detection method for most tests. But synchronous detectors are phase sensitive.

To illustrate these two demodulation techniques let us take an example of an initial signal in the form of two components (Figure 3.3a). Their composition is shown in Figure 3.3b. The HT envelope of the composition is a fast-varying function, but the synchronous demodulated amplitude precisely extracts a correspondingly fast component (Figure 3.3c). For a synchronous demodulation we should know *a priori* the frequency of the desired fast component.

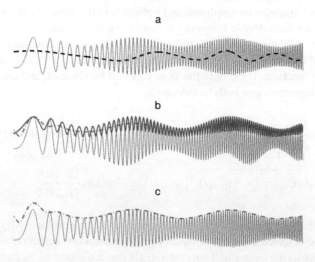

Figure 3.3   The HT and the synchronous demodulations: two components of the composition (a), the composition (b, —), the HT envelope (b, —), the synchronous amplitude (b, ⋯), the demodulated component (c, —), and the synchronous amplitude (c, ⋯) (Feldman, ©2011 by Elsevier)

A coherent detection is well recognized as a method of obtaining a quantum-limited reception in spectral ranges where a direct detection cannot be made quantum-limited. However, even in an ideal case, the signal-to-noise ratio of a coherent receiver is limited by more than just the quantum ef ciency of the detector (Fink, 1966).

3.3   Digital Hilbert transformers

Modern signal processing is almost digital signal processing, involving the use of digital procedures and signals digitized with the sampling frequency F_s. The HT is a procedure used to generate a quadrature component of a detrended real-valued "analytic-like" signal in order to analyze variations of the instantaneous phase and amplitude. We cannot design a digital HT procedure with a response function corresponding to either the ideal impulse response $(\pi t)^{-1}$ or the frequency response $H(\omega) = -i\,\mathrm{sgn}(\omega)$ (Figure 2.1). This frequency response describes an ideal wideband $90°$ phase shifter whose positive frequencies are shifted by $-90°$ $\left(-\frac{\pi}{2}\right)$ and whose negative frequencies are shifted by $+90°$ $\left(\frac{\pi}{2}\right)$. For example, for a signal with the form $x(t) = \sum\limits_{l=1}^{\infty} a_l \cos(l\omega t + \varphi_l)$ the Hilbert transformer should produce $H[x(t)] = \sum\limits_{l=1}^{\infty} a_l \cos(l\omega t + \varphi_l - 90°)$. The ideal Hilbert transformer, or phase shifter, which affects a $90°$ phase shift at all frequencies, cannot be realized perfectly. A practical lter differs from an ideal lter because it has loss characteristics. There are mainly two methods for obtaining a good approximate Hilbert transformer: frequency domain and time domain. They both synthesize an imaginary component of a complex analytic waveform from the real signal projection. The real component should be unchanged. Any algorithm for the HT realization is only an approximation. It may work well over a certain limited (mostly central) part of the frequency band, but it will not work well near the band ends.

3.3.1   Frequency domain

The Hilbert transformer may be implemented ef ciently using the fast Fourier transform. Following Fourier transformation, the negative frequencies are zeroed. An inverse Fourier transform will then yield a $90°$ phase-shifted version of the original waveform. A frequency domain technique is based on computing the Fourier transform of a signal and is an ef cient structure for implementing the HT (Claerbout, 1976). If $x(t)$ is a real input data record of a length N, then the analytic signal $X(t) = x(t) + i\tilde{x}(t)$ can be obtained by: $X(t) = \mathrm{IFFT}\{B(n) \cdot \mathrm{FFT}[x(t)]\}$ where: $B(n) = 2$ for $n = \{0, N/2 - 1\}$; $B(n) = 0$ for $n = \{N/2, N - 1\}$; FFT is the Fourier transform; and IFFT is the inverse Fourier transform. The imaginary portion of $X(t)$ contains the HT projection $\tilde{x}(t)$; the real portion contains the real input signal $x(t)$. Windowing and/or zero padding may have to be used to avoid ringing. For example, the MATLAB® procedure "hilbert.m" uses an FFT approach and the IFFT then produces a complex analytic waveform. The problems with this approach are the same as with any FFT technique that operates in a given frequency band and can suffer the effects of truncation.

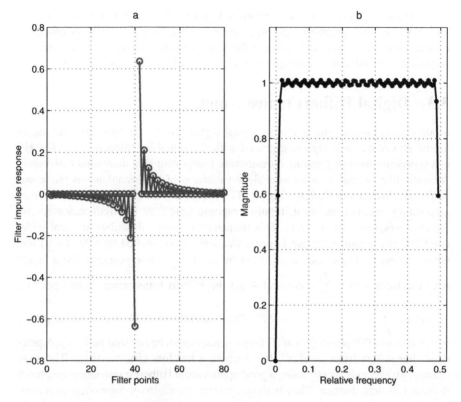

Figure 3.4   The digital Hilbert transformer as a filter: the impulse characteristics (a); the magnitude (b) (Feldman, ©2011 by Elsevier)

3.3.2   Time domain

An alternative approach is to synthesize the imaginary component directly from the real component using a time-domain lter (Lyons, 2000). A standard linear estimation (or prediction) technique may be employed to obtain a Finite Impulse Response (FIR) digital lter of a speci ed length that best approximates the Hilbert (Figure 3.4a) (Mitra and Kaiser, 1993). FIR digital Hilbert transformers based on the Remez exchange algorithm have an advantage of providing an exact linear phase. However, to achieve a given negative frequency attenuation level, they also require a higher lter order compared to the in nite impulse response (IIR) designs. A phase shift can be implemented by a lter, obtainable by the convolution theorem relating convolution to multiplication in the frequency domain. A limitation of the FIR digital Hilbert transformers applies to the length of lters because the design algorithm encounters numerical problems with large lter lengths (more than 200).

Recently C. Turner suggested a new Hilbert transformer by constructing a pair of quasilinear phase bandpass lters that have identical magnitude responses, differ in phase by 90°, and can be used for analytic signal generation (Turner, 2009). The lter has useful symmetry properties that signi cantly reduce their computational complexity and coef cient storage requirements, and there is no inherent length limitation to the size of the lter. Turner's prescription for the shifter includes taking an input signal and splitting it into two copies. Feeding one through the Hilbert transformer and the other through a delay (the delay equals the average delay of the HT) produces two components that now represent the analytic signal.

Any digital Hilbert transformer operates inside its frequency band limits (Figure 3.4b). It does not perform the HT at low frequencies (close to 0 Hz), nor at high frequencies close to the Nyquist frequency (Fs/2). Inside its passbands the lter has a rather at magnitude response (Figure 3.4b). For example, the MATLAB procedure " rpm ([], 'Hilbert')" uses the FIR lter approach with a quasilinear phase to produce the HT. By choosing a large value of order (length) for the lter, we can minimize the gain errors of the magnitude and phase responses.

The inherent limitations of the HT in the frequency domain determine the minimum and maximum number of samples during one period of signal oscillation. A low cutoff frequency $f_{min} \approx 0.02 Fs$ indicates the lowest frequency suitable for the HT. Thus, the corresponding largest number of samples per period will be $n_{max} = \frac{2Fs}{f_{min}} \leq 100$. A high cutoff frequency $f_{max} \approx 0.48 Fs$ indicates the maximum frequency suitable for the HT. Thus, the corresponding minimum number of samples per period will be $n_{min} = \frac{2Fs}{f_{max}} \geq 4$.

Different methods for a computation of the HT in the time domain are presented and discussed in Veltcheva, Cavaco, and Soares (2003). These methods depend on a concrete realization of the moving average procedure and show dissimilar precision in the envelope calculation.

3.4   Instantaneous characteristics distortions

A gain error is not the main source of digital inaccuracy with the HT. A degree of distortion injected by ltering within the frequency band limits may be quite considerable (Caciotta *et al.*, 2009). Distortions and errors of the Hilbert-transformed waveform depend on the shape of the initial signal. The maximum output distortion of the HT projection will be induced by step-and-impulse type variations in the initial signal. An example of step change of the signal amplitude is shown in Figure 3.5a.

The HT lter is transforming the instant amplitude step for the envelope function with some inertia. The resultant transient envelope has characteristics of the second-order step response, like decaying oscillations (Figure 3.5a). It can be seen that the estimated envelope returns to the steady condition only after several complete cycles of vibration. The instant frequency step is converted in the IF, also with some inertia. But due to the inherent differentiation procedure, the IF will have a more complicated transient form (Figure 3.5b). The estimated envelope returns to the steady condition after several complete cycles of vibration.

The HT lter is sensitive to short pulse errors. Even a single false spike propagates through the calculation of the instantaneous parameters and performs a serious transient distortion in the envelope and the IF (Figure 3.5c). It is clear that random noise, as an addition to the pure signal, also produces a serious distortion in instantaneous functions (Vakman, 2000) (Figure 3.5d).

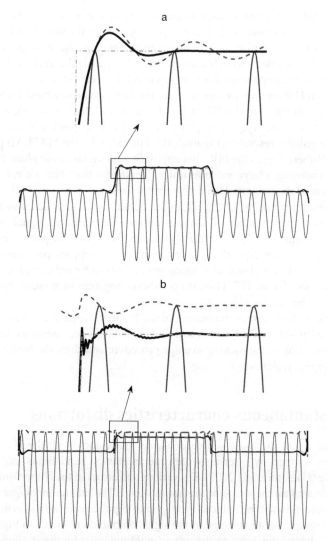

Figure 3.5 The distortion of the envelope and the IF: the amplitude step (a), the frequency step (b), the single spike (c), the random noise (d); the signal (—), the envelope (– –), the IF (—), and the noise (. . .) (Feldman, ©2011 by Elsevier)

3.4.1   Total harmonic distortion and noise

A signal analysis often deals with many different types of distortion, but the most common are harmonic distortions and random noise. A total harmonic distortion (a coef cient of harmonics, a coef cient of nonlinear distortions, a distortion factor) is a ratio of the sum of the powers of all harmonic frequencies, different from the fundamental frequency, plus a noise to the power of the fundamental

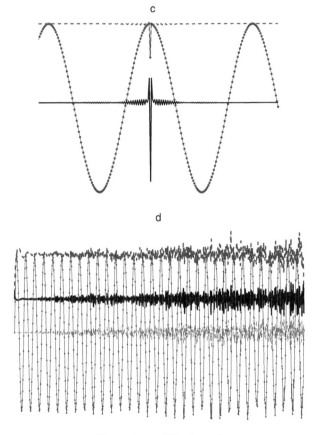

Figure 3.5   (Continued)

frequency: THD $= \dfrac{\Sigma \text{harmonics powers} + \text{noise power}}{\text{fundamental frequency power}}$. Obviously, if no other non-linear harmonics or noise are present, then the total harmonic distortion (THD) is zero. In the case of an additive noise, the THD is a *ratio* of an undesired *noise* to a desired *signal* in the average power level and provides a *noise to signal ratio*. Of course, the THD also characterizes the nonlinearity of the systems − such as a transient intermodulation distortion, an intermodulation distortion, and others. The HT is very sensitive to additive noises in the signal. It operates well only for small values of *noise to signal ratio* (Figure 3.5d)

3.4.2   End effect of the Hilbert transform

The end effect, or the Gibbs phenomenon, appears in a digital ltering or discrete Fourier transform due to an incomplete data periodicity, when the waveform has not completed a full cycle within its period of the analysis. The end effect problem exists for any digital analysis method. Traditionally, signal windows with tapered ends were used to alleviate these effects during the spectral analysis. In practice, however, the end effect of the HT is relatively easy to   x (Wu and Huang, 2009). We can also pad a signal using a segment of the signal itself in order to make the signal quasiperiodic (Huang *et al.*, 1998). After the construction of an arti cial padded data, we will compute the digital HT with a minimized end effect. We can use a data- ipping (mirror) technique to furnish a complete periodic cycle of the waveform to suppress the end effect (Huang, 2003). At least we can simply increase the initial data length to make the end effect negligibly small.

3.5   Conclusions

This chapter describes some modern digital-processing procedures for estimating the IF and instantaneous amplitude of real signals. There are two methods for obtaining the Hilbert transformer: the frequency domain and the time domain. In the frequency domain the Fast Fourier transform is used, and in the time domain the phase shift can be implemented by a convolution   lter, obtainable by the convolution theorem relating convolution to multiplication in the frequency domain. The performance of the HT-based estimation method is illustrated for noise and harmonic pollution sensitivity. The HT demodulation methods can also be employed to extract simple components using the varying instantaneous frequency and amplitude from multicomponent nonstationary signals.

Part II

HILBERT TRANSFORM AND VIBRATION SIGNALS

4

Typical examples and description of vibration data

The central problem in vibration analysis is a time series analysis. Spectral methods have been used as the standard tool for many years, but recently a serious interest has been shown in extending the analysis for the examination of a nonstationary time series. Both the spectral and the time domain methods have their strengths and weaknesses. It seems, however, that there is a compromise between resolution in time and frequency, and a corresponding magnitude of cross terms. For a general nonstationary vibration signal, the time–frequency and the analytic signal method do a better job of simultaneously localizing main signal components.

The search for methods to analyze signals, aimed at obtaining the attributes related to the physical properties that generate these signals, has always been a topic of interest. The analytic signal method is equally applicable to deterministic and random processes, although, generally speaking, it does not separate them at all, which is why it enables us to investigate any oscillating time function from a general point of view. The method is also good for solving problems concerning the analysis of stationary and nonstationary vibrations, as well as narrowband and/or wideband signals. It also allows a precise analysis of the transformation and dissipation of vibration energy and vibration effects on machine durability. This chapter lists and demonstrates some typical examples of mechanical vibration that can occur in practice, together with their instantaneous characteristics obtained through the application of the HT.

4.1 Random signal

As mentioned, the behavior of random signals can be analyzed using probability and statistical functions. For such signals both the envelope and the IF are also random functions (Figure 4.1). The wideband random vibration has a broad "white noise" type equal power spectrum (Figure 4.2a, dash line) whereas the narrowband random

Hilbert Transform Applications in Mechanical Vibration, First Edition. Michael Feldman.
© 2011 John Wiley & Sons, Ltd. Published 2011 by John Wiley & Sons, Ltd.

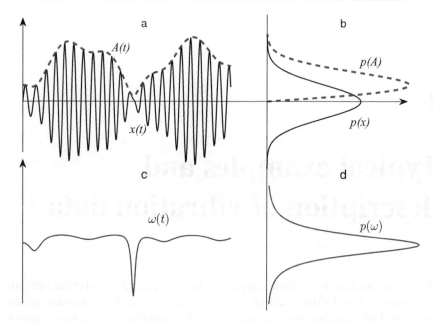

*Figure 4.1   The random signal and the envelope (a), the signal distribution (b, —),
the envelope distribution (b, – –), the IF (c), and its distribution (d) (Feldman, ©2011
by Elsevier)*

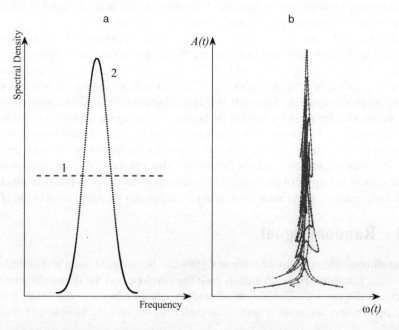

*Figure 4.2   The random signal spectrum (a): the wideband (– –) and narrowband
(—); the narrowband signal envelope vs. IF plot (b)*

Table 4.1   Typical examples of the central frequency and the spectral bandwidth of random vibration (Feldman, 2009b)

		Random vibration signal	
The IF spectral characteristics		Narrowband in the range $\omega_0 - \frac{1}{2}\Delta\omega < \omega < \omega_0 + \frac{1}{2}\Delta\omega$	System displacement $\ddot{x} + 2hx + \omega_0^2 x = F(t)$
Central frequency	Mean value $\bar{\omega}$	ω_0	$\omega_0 - 2h/\pi$
Spectral bandwidth	Mean modulus $\overline{\lvert\omega\rvert}$	$\omega_0(1 + \Delta\omega^2 24\omega_0^2)$	ω_0
	Energy equivalent $\Delta\omega_1$	$\Delta\omega$	πh
	IF deviation $\Delta\omega_2$	$\Delta\omega\sqrt{3}/6$	$2\left(h\omega_0/\pi\right)^{1/2}$

(Feldman, ©2009 by John Wiley & Sons, Ltd.)

vibration has a thin type power spectrum (Figure 4.2a, dotted line) concentrated around the central frequency (2.14).

The central frequency (2.14) and the frequency bandwidth (2.15), based on the IF distribution, are the fundamental characteristics of random vibration. Two typical examples of the central frequency and the spectral bandwidth estimations are shown in Table 4.1 for a narrowband random signal and for the random displacement on the output of the single-degree-of-freedom (SDOF) system under white noise force excitation (Bunimovich, 1951). The model of the vibration system has an undamped angular frequency ω_0, and damping factor (coefcient) $h = \zeta\omega_0$, where ζ is a constant called the *damping ratio* (see Section 8.8). The damping ratio also can be approximated from the *loss factor* η by the following formula (Inman, 1994), which is more accurate at lower damping: $\eta = 2\zeta$.

The energy of a narrowband signal is concentrated around a central frequency inside the signal's bandwidth. It is notable that, in the case of random vibration of the SDOF system, the central frequency $\bar{\omega}$ is less than the undamped angular frequency ω_0. It is even less than the damped free vibration natural frequency, which is equal to $\omega_{damped} = \left(\omega_0^2 - h^2\right)^{1/2} \approx \omega_0 - h^2/2\omega_0$, where h is the damping factor. In particular, the spectral bandwidth of random vibration of an SDOF system is proportional to the system's damping factor (see Table 4.1).

4.2   Decay vibration waveform

Such an exponentially damped sinusoid could be, for example, the impulse response of a linear SDOF system. The envelope of the signal is determined by a monotonic exponent decay rate. A free decay $x(t)$ is a well-known oscillation function with an amplitude gradually decreasing to zero (Figure 4.3) $x(t) = A_0 e^{-ht} \sin \omega_0 t$, where A_0

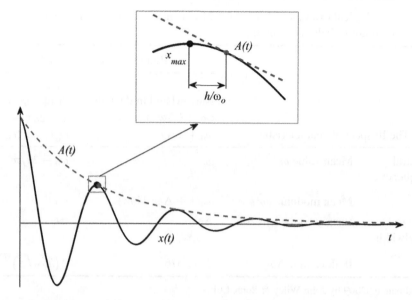

Figure 4.3   The damped oscillation (—) and the envelope (––). The envelope and the extrema points (Feldman, ©2011 by Elsevier)

is an initial amplitude, ω_0 is a frequency of the sinusoid, and h is the damping coefﬁcient (a measure of the amount of energy damping). Its time-dependent amplitude (the envelope) is the exponential function of time $A(t) = A_0 e^{-ht}$. The decreasing envelope touches the oscillation function at the points of contacts $t_k = (k\pi + h)/\omega_0$. It is signiﬁcant that these envelope contact points $A(t)$ do not correspond to the local extrema (peak points) $x_{\max}$ of the free decay function positioned at $t_{\text{peak}} = k\pi/\omega_0$ where $x = 0$ (Figure 4.3).

The ratio between two envelope values, taken at the initial and the terminal time points, has the form $A(t_0)/A(t_0 + \Delta t) = e^{-h(t_0 + \Delta t)}$. Taking the natural logarithms of both sides of the equation above gives a simple formula: $\ln[A(t_0)] - \ln[A(t_0 + \Delta t)] = -h(t_0 + \Delta t)$. This means that the natural logarithm of the amplitude ratio for any two envelope points separated in time is directly proportional to the time interval. This permits an estimation of the damping coefﬁcient (the rate of decay) $h = \Delta \ln(A)/\Delta t$. The damping coefﬁcient h is equal to the time needed for the free oscillation to decay to $1/e$ of its initial energy.

The spectrum of the free decay waveform (Figure 4.4a) and the envelope vs. IF frequency plot (Figure 4.4b) demonstrate a high concentration of vibrational energy around the central frequency ω_0. The narrower the vibration signal spectrum, the slower the decay of the oscillation signal and the smaller the damping coefﬁcient.

4.3   Slow linear sweeping frequency signal

A quasi-monoharmonic signal (chirp) with a slow linear increasing or decreasing sweeping frequency is a typical example of the slow frequency modulation

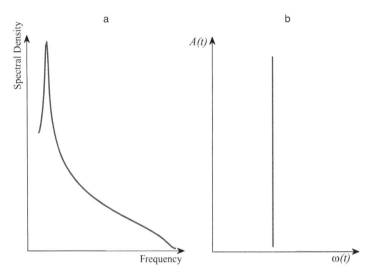

Figure 4.4   The damped oscillation: the spectrum (a), the envelope vs. IF plot (b)

(Figure 4.5) that is widely used for testing dynamics systems. For the case of linear frequency modulation, the IF of the carrier varies linearly with the modulating signal $\omega(t) = \omega_{\min} + k_\omega t$, where k_ω is a constant. The signal function in view of the integral instantaneous phase can be written as $x(t) = A_0 \cos\left[(\omega_{\min} + 1/2k_\omega t)\,t\right]$, where A_0 is a constant amplitude.

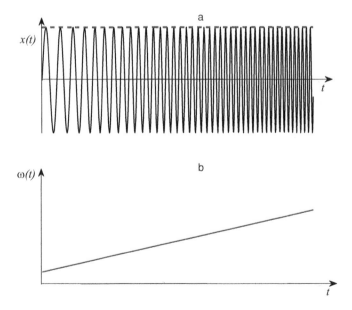

Figure 4.5   The frequency sweeping oscillation (—) and the envelope (a,--), the IF (b)

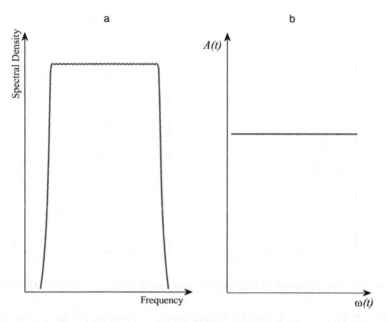

Figure 4.6   The frequency sweeping oscillation: the spectrum (a), the envelope vs. IF plot (b)

The power of a signal with constant amplitude does not vary in time and lies within a bandwidth of the carrier frequency $\Delta\omega = \omega_{max} - \omega_{min}$. During every short time interval it looks and acts like a harmonic. However, the total bandwidth $\Delta\omega$ of the signal that is swept through a large frequency range could be much more than its central frequency $\omega_0 = (\omega_{min} + \omega_{max})/2$, although the signal still behaves as a pure harmonic every time. A linear frequency sweeping has a white excitation spectrum (Figure 4.6a), while the exponential frequency sweeping has a pink excitation spectrum. The international organization for standardization published guidelines with ISO 7626 Part 2, "Measurements using single-point translation excitation with an attached vibration exciter," for the application of slowly swept sinusoids related to modal testing (Gloth and Sinapius, 2004).

4.4   Harmonic frequency modulation

When we modulate the frequency of a signal with a cosine wave modulating function $\omega_0 + \beta\omega_m \cos \omega_m t$ (Figure 4.7b), we will get a harmonic frequency modulated signal that varies in accordance with the modulating function (Figure 4.7a):

$$x(t) = A_0 \cos \int (\omega_0 + \beta\omega_m \cos \omega_m t)\, dt = A_0 \cos (\omega_0 t + \beta \sin \omega_m t), \qquad (4.1)$$

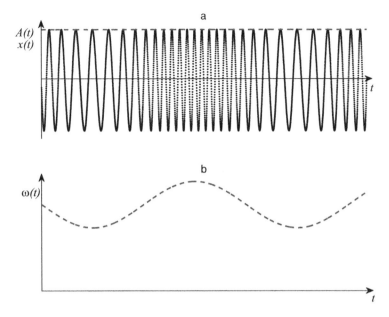

Figure 4.7   Harmonic frequency modulation: the signal (a), the signal IF (b)

where ω_0 is the carrier frequency, ω_m is the modulation frequency, and β is the frequency modulation index. The value $\beta\omega_m$ is the peak frequency deviation, which indicates the largest swing or deviation in the IF.

Rewriting Equation (4.1) by using Bessel series expansion (Cantrell, 2000) gives $x(t) = \sum_{-\infty}^{\infty} J_l(\beta) \cos[(\omega_0 + l\omega_m)t]$, where $J_l(\beta)$ is the Bessel function (rst kind, integer order l) for the β value. In essence, the last expansion is de ned as a spectrum of the fast frequency modulated signal. The spectrum consists of an in nite number of sidebands about the carrier frequency ω_0. A signal whose frequency is fast modulated with a harmonic has an nitive set of sidebands at frequencies $\omega_0 \pm |l|\,\omega_m$ that occur at multiples of the modulating frequency away from the carrier frequency (Figure 4.8a). With a larger frequency modulation index β we get more sidebands with larger amplitudes and a greater wideband spectrum.

The higher-order Bessel function values fall quickly with l when the modulation index is small. In practice when $|\beta| \ll 1$, $J_0(\beta) \approx 1$, $J_1(\beta) \approx 0.5$, and $J_{n\geq 2}(\beta) \approx 0$ the narrowband frequency-modulated spectrum can be approximated by only three members: $x(t) \approx J_0(\beta) \cos \omega_0 t + J_1(\beta)\{\cos[(\omega_0 - \omega_m)t] + \cos[(\omega_0 + \omega_m)t]\}$. Thus the time modulation parameter gives rise to additional spectral components.

It is interesting that the case of the fast-doubled modulation frequency $\omega_m = 2\omega_0$

$$x(t) = A_0 \cos(\omega_0 t + \beta \sin 2\omega_0 t) \approx [J_0(\beta) + J_1(\beta)] \cos \omega_0 t + J_1(\beta) \cos 3\omega_0 t$$
$$(4.2)$$

produces the carrier frequency and, at least, a sideband component as the tripled carrier frequency value. Generally, the fast frequency modulated signal combines a

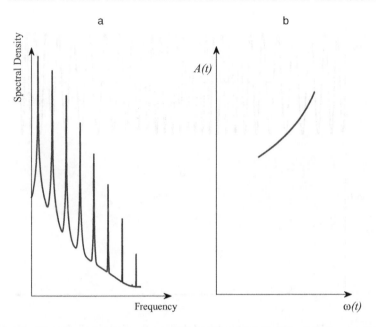

Figure 4.8   Fast harmonic frequency modulation: the spectrum (a), the envelope vs. IF plot (b)

number of harmonic components (Figure 4.8a). The IF of the related analytic signal is not equal to the initial frequency modulating function. The envelope of the fast frequency modulated signal does vary in time and the envelope vs. the IF has a form of a nonlinear curve (Figure 4.8b).

4.5   Harmonic amplitude modulation

Let us consider a harmonic amplitude modulation (AM) signal performed by modulating a cosine function on a carrier wave:

$$x(t) = (A_0 + A_m \cos \omega_m t) \cos \omega_0 t, \qquad (4.3)$$

where A_0 and ω_0 are the *unmodulated carrier amplitude* and the *carrier frequency* of the carrier; A_m and ω_m are, respectively, the modulation amplitude and the modulation frequency. The same AM signal can be rewritten in the form $x(t) = A_0 (1 + m \cos \omega_m t) \cos \omega_0 t$, where the steady signal $A_0 \cos(\omega_0 t)$ is called the *carrier*, the slow-varying multiplier $1 + m \cos \omega_m t$ is called the amplitude modulation function (the waveform that should be transmitted), and $m = A_m / A_0$ is the *modulation index* or *modulation depth*.

From the spectral structure the AM signal is composed of three mono-harmonics, each with different frequency. To nd the frequency spectrum, formula (4.3) can be rewritten in the form of a sum of constant amplitude signals by using a simple trigonometry manipulation

$$x(t) = A_0 \cos \omega_0 t + \frac{A_m}{2} \cos (\omega_m - \omega_0) t + \frac{A_m}{2} \cos (\omega_m + \omega_0) t. \qquad (4.4)$$

The sum indeed displays three separated spectral components (the carrier frequency ω_0 and two adjacent sidebands $\omega_0 - \omega_m$, $\omega_0 + \omega_m$). The power of an AM signal is concentrated at the carrier frequency and in two adjacent sidebands. Each sideband is equal in bandwidth and amplitude to that of the modulating signal. The AM that results in two sidebands and a carrier is often called a *double sideband amplitude modulation*. We always assume that the AM function multiplying a high-frequency carrier is a low-frequency function: $\omega_m \ll \omega_0$. Because of this, two adjacent sidebands are always positioned on the spectrum very close to the carrier frequency at an equal distance from the left and right sides. But despite the presence of a number of frequency components in the signal, the IF does not vary in time (see (4.5)) because both sidebands are equally spaced and each is a mirror image of the other. AM signals can differ by the modulation function or the value of the modulation indexes; however, as their IF does not depend on modulation, it will always be equal to the carrier frequency ω_0.

4.5.1   Envelope and instantaneous frequency of AM signal

The envelope of the modulation signal can be simply derived from either the direct form (4.3) or the composition form of three harmonic components (4.4) with the same trivial result: $A(t) = \pm\sqrt{x^2(t) + \tilde{x}^2(t)} = \pm A_m \cos \omega_m t$, where $\tilde{x}(t) = (A_0 + A_m \cos \omega_m t) \sin \omega_0 t$ is the HT projection according to the Bedrosian identity (2.2). The obtained envelope thus repeats the initial slow-varying amplitude modulation function of the signal.

The IF of the modulation signal also can be derived either from the direct form (4.3) or from the composition form of three harmonic components (4.4):

$$\omega(t) = d\{\arctan[\tilde{x}(t)/x(t)]\}\, dt = \omega_0. \qquad (4.5)$$

The obtained IF thus repeats the initial unmodulated carrier frequency, which is to say that AM has no influence on the IF of the signal. The carrier frequency is always positioned between the equally distanced sidebands, therefore the IF equal to the carrier frequency will be always located in the middle of the sidebands of the AM signal.

The initial vibration signal $X(t) = x(t) + i\tilde{x}(t) = A(t)e^{i[\varphi(t)+\omega_0 t]}$ can also be represented in another form of the analytic signal $X(t) = \left[A(t)e^{i\varphi(t)}e^{-i\omega_0 t}\right]e^{i\omega_0 t} = A_{compl}(t)e^{i\omega_0 t}$, where $A_{compl}(t) = A(t)e^{i\varphi(t)}e^{-i\omega_0 t}$ is the signal's *complex envelope* (see Section 4.5.3). The complex envelope is not unique because it is determined by an arbitrary ω_0 assignment. The spectrum of the complex envelope can be obtained by shifting an initial signal spectrum to the left toward the origin of axes. For narrowband signals the complex envelope varies slowly in time, and the complex envelope spectrum is centered around zero, and not the carrier frequency. The complex envelope has meaning mostly for overmodulated signals with a high modulation index when the oscillating envelope becomes negative.

4.5.2 Low modulation index

It is often assumed that the modulation index is less than 1 ($m = A_m/A_0 < 1$), resulting in a pure harmonic variation of the upper positive envelope $+A(t)$. For such a small depth of modulation, both the upper positive and the lower negative envelopes $-A(t)$ are anti-phased functions of time (Figure 4.9).

As we already know, the spectrum of the AM signal has three spectral peaks – the carrier frequency ω_0 and two adjacent sidebands $\omega_m - \omega_0$, $\omega_m + \omega_0$. The amplitude (power) of the central carrier frequency peak is higher than the equal amplitude of small sideband peaks. In the case of the constant carrier frequency, the IF takes the form of a vertical straight line (Figure 4.10).

4.5.3 High modulation index

Let us again take the AM signal modulated by a cosine wave, but with a modulation index greater than unity $m = A_m/A_0 > 1$. The carrier wave becomes overmodulated and the AM function $1 + m \cos \omega_m t$ will have negative values. As a result, some amplitude distortions will occur and the envelope itself will be changing much faster (Figure 4.11). This will result in the appearance of complicated riding cycles with local negative maxima or local positive minima in the signal waveform.

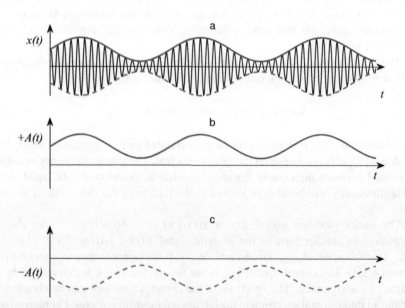

Figure 4.9   An AM with low modulation index: the signal (a), the upper positive envelope (b), the lower negative envelope (c)

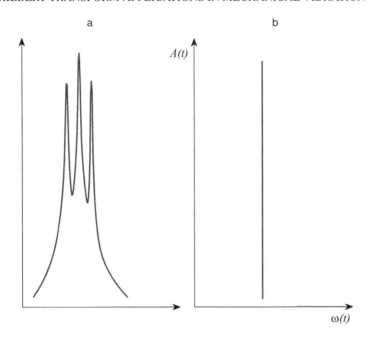

Figure 4.10   An AM with low modulation index: the spectrum (a), the envelope vs. IF plot (b)

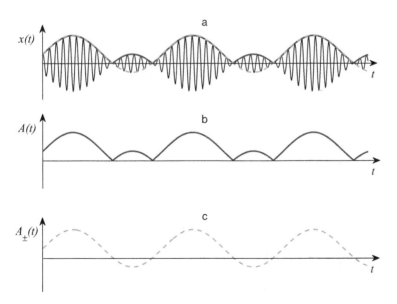

Figure 4.11   An AM with high modulation index: the signal (a), the upper positive envelope (b), the alternate envelope (c)

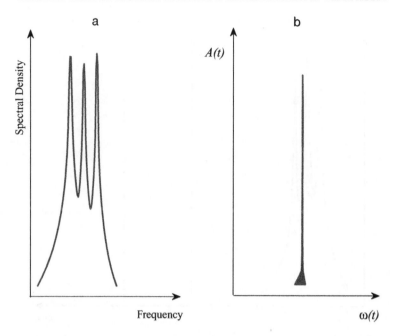

Figure 4.12   An AM with high modulation index: the spectrum (a), the envelope vs. IF plot (b)

The case of over-modulation generates duality relations for the envelope. Actually, on the one hand, the AM function $1 + m\cos\omega_m t$, having both positive and negative values (Figure 4.11(c)), can be described as the complex or *alternate envelope* $A_{\pm}(t)$ with positive and negative square root values (2.6). On the other hand, the envelope can be taken only with the positive sign (Figure 4.11b). If there are only non-negative values of the envelope in an overmodulated signal, signi cant phase discontinuities will be introduced. The choice of envelope representation between two possible forms depends on the operability and simplicity of further operations with the signal (Cohen, Loughlin, and Vakman, 1999; Cohen and Loughlin, 2003).

Figure 4.12a shows the spectrum in which we can see at once that the over-modulated signal has the same three components – a carrier wave and two sinusoidal sidebands – whose frequencies are slightly above and below. But the amplitude of the central peak of the carrier wave can be even less than the amplitude of the sideband waves. Over-modulation has no in uence on the IF of the signal (Figure 4.12b).

4.6   Product of two harmonics

The product of two harmonics is a particular case of an overmodulated signal when the carrier amplitude is $A_0 = 0$: $x(t) = A_m\cos\omega_m t\ \cos\omega_0 t$ (Figure 4.13a). The alternate envelope will be a pure harmonic function $A(t) = A_m\cos\omega_m t$ and the spectrum will have only two sideband components without a central peak: $x(t) = 0.5\,[\cos(\omega_0 - \omega_m)\,t + \cos(\omega_0 + \omega_m)\,t]$ (Figure 4.14a). This gure shows the

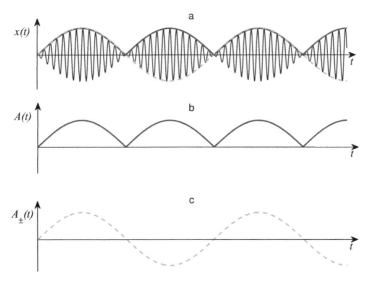

Figure 4.13   The product of two harmonics: the signal (a), the upper positive enve-
lope (b), the alternate envelope (c)

waveform variation of two pure tones with slightly different frequencies but the same
amplitudes.

It is interesting to note that, despite the absence of the carrier frequency in the
spectrum, the IF still is equal to the central frequency of the carrier, as the mean
value of the sideband frequencies (Figure 4.14b). Also, it is interesting that the case

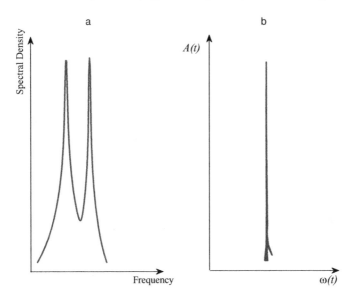

Figure 4.14   The product of two harmonics: the spectrum (a), the envelope vs. IF
plot (b)

of a sum of two harmonics with different arbitrary frequencies, but equal amplitudes, can be considered as the product of two harmonics. In other words, a composition of two harmonics with the same amplitude can always be expressed as a multiplication of the slow alternate envelope, with the frequency as half of the difference, and the carrier as half of the sum, of the frequencies of harmonics $\cos \omega_1 t + \cos \omega_2 t = 2 \cos [(\omega_2 - \omega_1) t/2] \cos [(\omega_2 + \omega_1) t/2]$. The IF of a composition of two harmonics with the same amplitude is equal to the mean value of their frequencies (Loughlin and Tacer, 1997; Suzuki *et al.*, 2006).

4.7 Single harmonic with DC offset

Let us consider the composition of a harmonic and a constant or slow-varying ape-riodic trend: $x(t) = A_0 \cos \omega_0 t + a$, where A_0 is the amplitude, ω_0 is the frequency of the harmonic, and a is a constant or the slow-varying trend (Figure 4.15). The DC offset (distortion) a is the time average value of the signal $a = (t_2 - t_1)^{-1} \int_{t_1}^{t_2} x(t)dt$. The HT projection of the signal looks even simpler than the initial signal because the HT of the constant is equal to zero: $\tilde{x}(t) = A_0 \sin \omega_0 t$. These two functions will produce an analytic signal with the oscillating envelope,

$$A(t) = \left[A_0^2 + a^2 + 2a A_0 \cos(\omega_0 t) \right]^{1/2}, \tag{4.6}$$

and with the IF varying in time

$$\omega(t) = \frac{A_0 \omega_0 (A_0 + a \cos \omega_0 t)}{A_0^2 + 2A_0 a \cos \omega_0 t + a^2}. \tag{4.7}$$

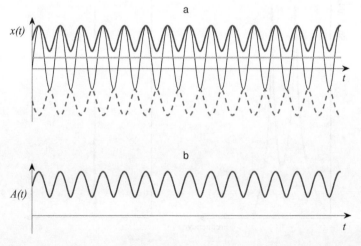

Figure 4.15 A harmonic with DC offset: the signal (a) with the envelopes, the constant DC; the upper positive envelope (b)

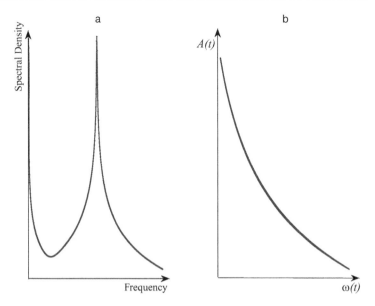

Figure 4.16   A harmonic with DC offset: the spectrum (a), the envelope vs. IF plot (b)

At rst glance the obtained IF signi cantly complicates the HT representation of the harmonic signal with the DC distortion (Figure 4.15). Really, a form of the IF now depends on the value of constant a. Note that in spectral analysis the same DC distortion will also initiate a large power spectrum peak on zero frequency, dependent on the constant a (Figure 4.16a).

Quite recently this example of the IF was used to illustrate an ostensibly serious problem of the HT to compute a physically valid value (Huang *et al.*, 2009). But in reality there are no problems with such a dependency, or with such possible negative values of the IF.

Signal phasor can be considered as a planar vector rotating around its center. Making a constant addition to the phasor projection means shifting (translating) the center of the vector rotation (the beginning of the rotated vector) from the origin to a new constant position along the horizontal axis. Relative to the new position, the vector describes the same circular orbit with the same angular velocity. But relative to the origin the vector will demonstrate a completely different and more complicated motion whose velocity depends on the constant of translation.

The same is true for the case of the sum of two rotating vectors. Two vectors added together produce a third resultant vector by placing the beginning of vector x_2 at the end of vector x_1. The vector sum $x_1 + x_2$ can be drawn as the vector from the origin to the end point (Figure 4.17). The angle of the resultant vector $x_1 + x_2$ can be found using the known trigonometric relations, while again the phase and the IF will depend not only on the phases, but also on the amplitudes of the components. Relative to the new point, the vector x_2 describes the same circular orbit and angular

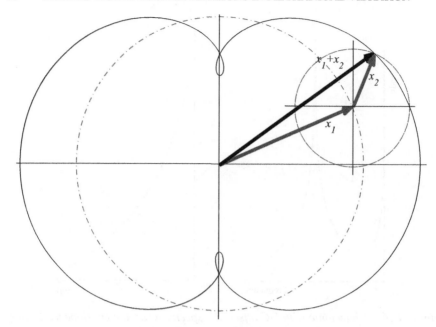

Figure 4.17   The relative and absolute motion of the sum of two vectors

velocity ω_2 will stay invariant with respect to the new choice of reference point. But relative to the origin, we can have a completely different motion where the IF will depend also on the vector amplitudes.

In kinematics, as a part of mechanics, it is a trivial situation of the planar motion. An object may appear to have one motion to one observer and a different motion to a second observer. There are no dif culties encountered here. To know the partial pure rotation we must subtract the new origin from the motion, and then perform an estimation of the angular frequency. Removal of the DC offset from the composition can easily be provided by different techniques, including the HT decomposition methods (Huang *et al.*, 1998; Feldman, 2006). Commonly in HT analysis, as well as in spectral analysis, we will consider the signal in which the aperiodic distortion has already been removed. This will return the envelope and the IF of a pure harmonic to the initial simplest constant values.

4.8   Composition of two harmonics

The superposition of two pure tones of different frequencies ($\omega_1 \neq \omega_2$) and amplitudes ($A_1 \neq A_2$) is shown in Figure 4.18, where two waves during some time interval are "in phase" and during some other interval are "out of phase". During the rst time interval there is a constructive interference, in which the amplitudes of the two waves add to make a wave with twice the amplitude of the individual pure tones. During the other time interval two waves produce a destructive

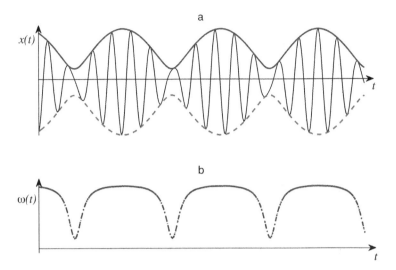

Figure 4.18   Two tones: the signal (a) with the envelopes, the IF (b)

interference, which gives a very small amplitude for the resulting wave formed by the superposition. For the signal composition as a sum of two harmonics: $x(t) = A_1 \cos \omega_1 t + A_2 \cos \omega_2 t$, $A_1 \neq A_2$, the envelope $A(t)$ of the double-component signal composition, according to Equation (2.6), could be written as:

$$A(t) = \left[A_1^2 + A_2^2 + 2A_1 A_2 \cos (\omega_2 - \omega_1) t \right]^{1/2}. \tag{4.8}$$

The signal envelope $A(t)$ consists of two different parts – a slow-varying part including the sum of the component amplitudes squared, and a rapidly varying part, oscillating with a new frequency equal to the difference between the component frequencies.

The IF $\omega(t)$ of the double-component composition according to (2.9) ($A_1 \neq A_2$) is:

$$\omega(t) = \omega_1 + \frac{(\omega_2 - \omega_1) \left[A_2^2 + A_1 A_2 \cos (\omega_2 - \omega_1) t \right]}{A^2(t)} \tag{4.9}$$

The IF of the two tones considered in Equation (4.9) is generally time-varying and exhibits asymmetrical deviations about the frequency ω_1 of the largest harmonic. The IF for two tones does not only have time-varying deviations, but these deviations always force the IF beyond the frequency range of the signal components. For large amplitude values of the second harmonic when

$$\frac{A_2}{A_1} > \frac{\omega_1}{\omega_2}, \tag{4.10}$$

the IF of the composition becomes negative. The appearance of the negative IF corresponds to the arrival of the local negative maximum or local positive minimum of the signal. The upper tangent to the negative maximum touches the signal negative envelope and, vice versa, the lower tangent to the positive minimum touches the signal positive envelope.The IF consists of two different parts: the frequency of the rst largest component ω_1 and a rapidly varying asymmetrical oscillating part. By eliminating the oscillating part $A_1 A_2 \cos(\omega_2 - \omega_1)t$ from Equations (4.8) and (4.9), we will receive an expression between the signal instantaneous characteristics (envelope vs. IF) as a function of the initial four parameters of the signal components:

$$A^2(t) = \frac{(A_1^2 - A_2^2)(\omega_2 - \omega_1)}{\omega_1 + \omega_2 - 2\omega(t)}, \quad A_1 \neq A_2, \quad \omega_1 \neq \omega_2. \tag{4.11}$$

Equation (4.11) determines the signal envelope as a function of the IF in the form of a hyperbola (see Figure 4.19b), whose length, direction, and curvature depend on four initial parameters of the initial bi-harmonic signal.

In the particular case of equality of the harmonic amplitudes ($A_1 = A_2$), when the IF of the signal is equal to the half of sum of component frequencies ($\omega(t) = (\omega_1 + \omega_2)/2$) (Suzuki et al., 2006; Loughlin and Tacer, 1997), this plot takes the form of a straight vertical line. The superposition of two pure tones of equal amplitudes and slightly different frequencies can be presented as a product of two harmonics (see an example above) with a fast carrier and modulated amplitude. If the frequency difference between the two pure tones is small enough, we can hear the loudness of the superposed wave varying with time. This loudness variation is known as *beats*. We hear a pure tone with a pitch given by half of the sum of the component frequencies. The amplitude variation occurs at the beat frequency, given by the difference between the two pure tone frequencies: $\omega_{beats} = \omega_2 - \omega_1$.

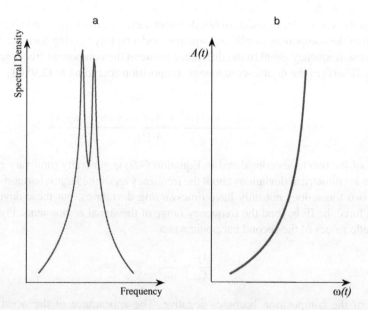

Figure 4.19   Two tones: the spectrum (a), the envelope vs. IF plot (b)

4.9   Derivative and integral of the analytic signal

In physics, the rst derivative of the displacement is equal to the velocity, and the second derivative, with respect to time, is the acceleration. It is also well known that the rst derivative (differentiating the function once with respect to time) yields the slope of the tangent to that function. Differentiating the analytic signal in the form $X(t) = A(t)e^{i\psi(t)}$ results in a relationship between an initial complex signal and its rst-order derivative as

$$X = A(t)e^{i\psi(t)} + i\omega(t)A(t)e^{i\psi(t)} = X\left[\frac{A}{A} + i\omega\right], \qquad (4.12)$$

where the IF $\omega = \dot{\psi}$ is the rst derivative of the instantaneous phase. The analytic signal notion expresses the rst derivative in terms of the envelope, its derivative, and the IF (4.12). The obtained expression for the rst derivative is carried by two components – the IF function as the angular velocity and the variation of the envelope as the radial velocity of the phasor (see Section 2.7).

The second derivative of the signal can be also expressed in analytic signal form:

$$\ddot{X} = X\left(\frac{\ddot{A}}{A} - \omega^2 + 2i\frac{A}{A}\omega + i\dot{\omega}\right), \qquad (4.13)$$

where the rst derivative of the IF $\omega = \dot{\psi}$ is the second derivative of the instantaneous phase. In the simplest case of a pure harmonic with a constant amplitude and frequency when $A = \ddot{A} = \dot{\omega} = 0$, we will have $\dot{X} = iX\omega$ and $\ddot{X} = -X\omega^2$.

Integration of vibration signals – as the opposite to differentiation – may also be of importance for signal processing. A complex function of a real variable t is an integral of the analytic signal if the real and the imaginary parts of the function are corresponding integrals forming a pair in the HT : $X(t) = \int_{t_1}^{t_2} X(t)dt = \int_{t_1}^{t_2} x(t)dt + i\int_{t_1}^{t_2} \tilde{x}(t)dt$. The integrated analytic signal as a de nite integral includes the integration constant. This means, for example, that the displacement resulting from the integration of the velocity will have an arbitrary DC offset. Theoretically, if the initial and nal conditions are known, it can minimize the overall integration error by increasing the accuracy at the boundaries, but in practice we mostly operate with an alternating current (AC) vibration signal without the offset. So the result of integration always will be the AC vibration without an offset value. To achieve a zero offset, it is a common practice to use highpass ltering after digital integration.

4.10   Signal level

The vibration level characterizes an intensity of alternating uctuations of the oscillating parameters – displacement, velocity, or acceleration. Vibration is often described as a random variable requiring an integral estimation approach. Thus the overall vibration level shows the spread (variability) of the signal around its average value which, for the most part, is zero.

4.10.1   Amplitude overall level

By using the envelope demodulation technique for a vibration signal in the time domain, we can nd the overall vibration level through the extracted envelope function. After the envelope of the signal is known, we can then make the classic direct estimations of the signal level by the envelope mean value $\overline{A} = T^{-1} \int_0^T A(t)dt$ and the envelope variation $\overline{[A(t)]^2} = T^{-1} \int_0^T A^2(t)dt$ (see also Section 2.3).

Although the envelope is not always known, we can nevertheless – even without the envelope demodulation – still approximately estimate the amplitude average value purely on the basis of the signal's energy. The analytic signal representation enables us to consider a vibration signal at any moment of time as a quasiharmonic wave, amplitude and phase modulated by time-varying functions $A(t)$ and $\psi(t)$: $x(t) = A(t) \cos \psi(t)$. For random signals, the envelope and the phase are statistically independent random functions. Thus, the signal is described as the product of two statistically independent functions: the envelope and the fast-oscillating cosine function. The variance of the product of these two functions can be written as: $\text{RMS}_x^2 = \left[\left(\overline{A} \right)^2 + \text{RMS}_A^2 \right] / 2$, where $\overline{A}$ is the envelope mean value , RMS_A is the envelope standard deviation , and RMS_x is the signal standard deviation. From the last expression we can rewrite the mean value of the envelope in the form of a function of the signal and envelope standard deviations: $\overline{A} = \left(2 \times \text{RMS}_x^2 - \text{RMS}_A^2 \right)^{\frac{1}{2}}$. In practice the signal standard deviation RMS_x^2, as the signal energy can be easily estimated. That cannot be said about the envelope standard deviations RMS_A^2, which are equal to zero for a pure harmonic or to half of the signal energy for a random signal. Considering these two opposite extreme cases of the harmonic and random signals, we will get a rather close values for an average envelope: $\overline{A}_{\text{random}} = \text{RMS}_x \sqrt{\pi/2} \approx \text{RMS}_x \times 1.25$; $\overline{A}_{\text{harmonic}} = \text{RMS}_x \sqrt{2} \approx \text{RMS}_x \times 1.41$; where the difference is only 10%. This means that the average envelope value in an arbitrary case between a pure harmonic and a random signal lies between these two close estimations of the signal standard deviation $\overline{A} \approx (1.25 \div 1.41) \text{RMS}_x$.

The initial signal and its envelope have common tangents at points of contact, but the signal never crosses the envelope. The common points of contact between the signal and its envelope do not always correspond to the local extrema of a multicomponent signal (see the example in Figure 4.3). The local extrema always have a zero tangent slope, but the common points of the contact can have a nonzero value of the tangent slope. Every *maximum* (top extremum) and *minimum* (bottom extremum) point of a function, known collectively as the set of local extrema points x_{extr}, is uniquely de ned by the rst derivative when the slope of the tangent is equal to zero. The distance between the common envelope points and the signal extrema points plays a dominant role in the theoretical explanation of the Empirical Mode Decomposition (EMD) mechanism (Huang *et al.*, 1998). In effect, without such a difference between the envelope and the local extrema, the sum of maxima and minima curves required by the EMD would always be equal to zero, just like the zero sum of the upper and the lower envelopes (see Section 6.2).

4.10.2   Amplitude local level

Traditionally, the analysis of deterministic vibrations uses the concept of the peak value as the absolute value of the maximum or minimum of the oscillating parameter during the time segment. For a random vibration, the peak value characterizes only the quasimaximum level, above which the signal is possible within a certain probability. For a random normal vibration, the most common maximum amplitude holds the value between plus and minus three times the RMS value of the signal $x_{\text{max}/\text{min}} = \pm 3\text{RMS}$, known by its statistical name of three sigma peaks. Thus, 99.73% of values of the vibration signal fall within these de ned limits, and only 0.27% fall out of them. A very large data sample size can yield signal random peaks outside the $\pm 3\text{RMS}$ limits.

For deterministic vibration and, in particular, for a mono-harmonic signal, the peak value is equal to its amplitude, and the peak to peak magnitude is a double amplitude. Note that the peak amplitude found for digital data is approximate, since the true peak of the output sinusoid generally occurs between samples. We will nd the local level estimation more accurately in the next sections.

4.10.3   Points of contact between envelope and signal

Both the envelope $A(t)$ and the initial function $x(t)$ can be functions that vary over time. Their relation has a known and simple form: $x(t) = A(t)\cos[\varphi(t)]$. Varying, the initial function and its envelope will have common contact points where these functions touch each other. The function and the envelope have common tangents at these points of contact. The condition of their contact takes the form:

$$x(t) = \pm A(t), \;\; \cos[\varphi(t)] = \pm 1, \;\; \varphi(t) = \begin{cases} 0 \\ \pm \pi \end{cases}, \qquad (4.14)$$

indicating that the common contact points are always located on the envelope. In other words, the envelope of nonstationary signal does not always pass through its peak values.

4.10.4   Local extrema points

Unlike the common contact points, the local extrema depend on the zero slope of the rst derivative. The signal derivative can be expressed as

$$X(t) = \sqrt{A^2(t) + A^2(t)\omega(t)^2} \exp\left\{ i \left[\varphi(t) + \arctan \frac{A(t)\omega(t)}{A(t)} \right] \right\} \qquad (4.15)$$

It introduces a new varying velocity envelope and a new varying velocity phase function. Let us nd points at which the slope of the tangent is zero and at which the function reaches its extrema. These points correspond to the zero value of the cosine function of the new varying velocity phase angle (4.15):

$$\varphi(t) + \arctan \frac{A(t)\omega(t)}{A(t)} = \pm \frac{\pi}{2},$$

$$\varphi(t) = \pm \frac{\pi}{2} - \arctan \frac{A(t)\omega(t)}{A(t)}. \qquad (4.16)$$

Notice that in the case of a mono-harmonic signal, the envelope is constant ($A(t) = 0$) and the conditions in (4.14) and (4.16) become identical: $\varphi(t) = 0$, or $\varphi(t) = \pm\pi$. For a mono-harmonic, the local extrema always lie on the envelope.

4.10.5 Deviation of local extrema from envelope

In a general case, the vertical position of the local extrema $x_{extr}(t) = A(t)\cos[\varphi(t)]$ is determined by the cosine projection of the new velocity phase (4.15):

$$x_{extr}(t) = A(t)\cos[\varphi(t)] = \pm \frac{A^2(t)\omega(t)}{A(t)\sqrt{1 + \frac{A^2(t)\omega^2(t)}{A^2(t)}}} \qquad (4.17)$$

The obtained continuous vertical position of the local extrema differs from the envelope function. The distance between the envelope and the extrema depends on the cosine projection from (4.17). Generally, the cosine projection assumes values from 1 through 0 up to −1. Therefore the corresponding local maxima can be equal to the envelope value, be as small as zero, or even take negative values. The cosine projection is controlled by the variable $\frac{A(t)\omega(t)}{A(t)}$ whose shape, level, and rate depend on the initial signal $x(t)$. In turn, the variable $\frac{A(t)\omega(t)}{A(t)}$ is determined by the relation between the nominator and the denominator.

For small envelope variations, when $A_{max} \ll (A\omega)_{max}$, the value of the cosine projection $\cos[\varphi(t)]$ does not differ from unity. This condition always forces the local maxima to be on the envelope. So, connected local maxima and connected local minima curves just repeat the corresponding – with opposite signs – upper and lower envelopes. For larger envelope variations, the cosine projection during oscillation can decrease to zero, or even to negative values. It will produce zero and negative local maxima below zero up to the opposite signed lower envelope. An example of the distance between the envelope and the extrema for two harmonics is given in Section 6.2.3.

4.10.6 Local extrema sampling

Connected together, the local maxima give shape to the maxima curve and, correspondingly, connected local minima give shape to the minima curve. As shown in (4.17), the vertical values of both the top and bottom extrema curves are generated by the continuous function of multiplying the envelope by the cosine projection. However, the actual discrete extrema points themselves are formed by digitizing the continuous cosine projection at distinct moments of time. These sampling moments completely depend on the IF $\omega(t)$ of the initial signal. Thus, from the continuous function of the cosine projection $x_{extr}(t)$ (4.17) we have a set of extrema sampled with the IF $\omega(t)$. The series of sampled extrema interpolated by the spline generate the two continuous pro les (extrema curves) required by the EMD method (Huang *et al.*, 1998).

The discrete set of samples obtained does not repeat the original continuous function $x_{extr}(t)$. If the frequency of the continuous function (4.17) exceeds (overlaps) the Nyquist frequency $0.5\omega(t)$, the sampled extrema line undergoes aliasing with a new folding frequency of half of the sampling frequency: $\omega_{fold}(t) = |\omega(t) - \omega_{x_{extr}}(t)|$. But if the frequency of the continuous function lies below the Nyquist frequency $0.5\omega(t)$, no aliasing occurs, and the sampled extrema curves will follow the signal envelope. A case of the frequency limit distinguishing the closest harmonics is presented in Section 6.2.7.

4.11   Frequency contents

The traditional way to nd out the frequency composition of signals is to examine them in the frequency domain. This is how the Fourier spectrum shows the spread of signal energy over spectrum frequencies. The IF, taken as the derivative of the instantaneous phase (2.9), is something different because it describes the angular velocity rate of the analytic signal at every particular time. It allows us to know the frequency that exists at any moment and how it changes with time. The IF by itself is a useful analytic signal feature; in combination with the envelope it forms a valuable 2D envelope vs. IF plot (see Section 2.1) which reveals the relations between these instantaneous characteristics. Some simple examples of a typical envelope–frequency plot were discussed in Sections 4.1–4.8.

As we can see, the IF presents a clear frequency contents in simple cases of a narrow range of frequencies that slowly vary in time. These cases are restricted to the mono-component and narrow-band signal when the IF can be interpreted as the mean, or the median, frequency existing during a short time interval at the center of the narrow-band frequency distribution. In a more general case, when the IF becomes negative or exceeds the limits of the spectral frequencies, we need to nd a way to interpret the IF of such a multicomponent signal. The IF for a multicomponent signal is de ned as a global characteristic collectively described by the instantaneous frequencies associated with all the individual components in the signal. For the multicomponent signal it could be solved by considering a proper method of decomposition a complicated vibration into simpler components rst (Braun and Feldman, 2011).

4.12   Narrowband and wideband signals

Classically, the relations between the spectral bandwidth $\Delta\omega$ and the central frequency ω_0 allow vibration signals to be divided into two groups: narrow band $(\Delta\omega/\omega_0 \ll 1)$ and wide band $(\Delta\omega/\omega_0 \gg 1)$. This distinction of the signals is rather conventional, but it does not completely or correctly describe the real nature and the frequency band type of the signal. The relations between the spectral bandwidth $\Delta\omega$ and the central frequency ω_0 can be expressed in an analytic signal form. To realize it, the mean value of the IF $\bar{\omega}$ is considered as the central frequency of the signal $\bar{\omega} = \omega_0$ (2.14), and the instantaneous bandwidth of the signal $\Delta\omega$ as the sum of the IF and the envelope variations (2.15): $\Delta\omega^2 = \overline{\omega^2} + \overline{\dot{A}^2} - \bar{\omega}^2 = \sigma_\omega^2 + \overline{\dot{A}^2}$, where $\sigma_\omega^2 = \overline{\omega^2} - \bar{\omega}^2$ is the IF variance. The conventional condition $\Delta\omega/\bar{\omega}$ of breaking vibration down to narrow band or wide band is here rearranged to the following form: $\Delta\omega^2/\bar{\omega}^2 = \sigma_\omega^2/\bar{\omega}^2 + \overline{\dot{A}^2}/\bar{\omega}^2$. The condition obtained is expressed by the sum of two variance-to-mean ratios: the IF and the envelope relative variation ratios. It is assumed that the variance-to-mean ratio, like the coef cient of variation, is a normalized measure of the dispersion of a probability distribution. Smaller values of variance-to-mean ratios (<1.0) correspond to a more concentrated distribution in the narrow frequency band. Larger values of variance-to-mean ratios (>1.0) correspond to the existence of a wideband frequency vibration.

Summarizing, we can determine a signal frequency band as the sum of its IF and the relative envelope variations. A detrended (centered) vibration data taken during several cycles is considered to be a narrowband signal if its IF and/or the envelope variance-to-mean ratios are less than 1. A vibration signal with a larger IF and/or tenvelope variance-to-mean ratio is considered to be a wideband signal. Therefore a slow frequency modulated signal is a typical example of a narrowband signal. With this de nition in mind, the narrowband signal in each cycle, de ned by the central frequency, involves only one mode of oscillation; no complex riding waves are allowed (Huang, Shen, and Long, 1999). In effect, in this case the IF will not have the fast uctuations that other waveforms induce. The explicit expression for the spread of the IF and the envelope derivative at a particular time clari es the meaning of the narrow- and wideband signals.

4.13   Conclusions

Several typical waveforms are considered in this chapter. They are common in mechanical vibration measurements and excitation of the required structural motion. The most common types are harmonic modulated signals, including the swept sine, decay vibration waveform, and random signal. They all differ in their amplitude and the IF contents. For the monoharmonic signal, the peak value is equal to its amplitude, and the peak to peak magnitude is a double amplitude. However, in a general case the envelope of a nonstationary signal does not pass through its peak values.

The signal frequency band is determined as the sum of the signal IF and the relative envelope variations. Vibration taken during several cycles is considered to be a narrowband signal if its IF and/or the envelope variance-to-mean ratios are less than 1. A vibration signal with a larger IF and/or envelope variance-to-mean ratios is considered to be a wideband signal.

5

Actual signal contents

In addition to the signal bandwidth parameter, there is another very significant category of vibration. It characterizes a complexity of the signal – the existence of a number of simpler parts (components) of the signal that form a multipart signal content. Such a multipart signal composition has been called a multicomponent signal. The multicomponent signal, similar to any other arbitrary signal, has its envelope and the IF at any moment of time, but now these instantaneous functions become rather complicated functions of time. A simplest example of the multicomponent signal is a mixture of two or several harmonics (Section 4.8). Even this simplest example shows that the IF is widely spread without a local concentration and could not be associated with any of the signal parts (Loughlin and Tacer, 1997). So neither signal part is compact about its IF.

According to Putland and Boashash (2000), a monocomponent signal is one whose time–frequency distributions is a single "ridge" – that is, a single delineated region of energy concentration. It is also required that the ridge does not "fold back" in time; that is, interpreting the crest of the "ridge" as a graph of frequency vs. time, we require the frequency of a monocomponent signal to be single-valued. A multicomponent signal is one whose time–frequency distribution comprises two or more ridges, representing the sum of two or more monocomponent signals; a stationary signal is one whose time–frequency distribution is independent of time.

5.1 Monocomponent signal

A nonstationary signal can have a slow-varying amplitude and a slow-varying always positive IF. This kind of signal is often referred to as a single- (mono-) component, or monochromatic signal $x(t) = A(t) \cos \int \omega(t) dt$, $A(t) > 0$, $\omega(t) > 0$. Here a "monocomponent function" is used as a technical term and indicates an oscillating function close to the most common and basic elementary harmonic function. The term "monocomponent signal" is almost the same as the "intrinsic mode function"

Hilbert Transform Applications in Mechanical Vibration, First Edition. Michael Feldman.
© 2011 John Wiley & Sons, Ltd. Published 2011 by John Wiley & Sons, Ltd.

term suggested in (Huang *et al.*, 1998). While the IF is always positive, the signal itself has the same numbers of zero crossings and extrema. When the IF takes a negative value, the signal has multiple extrema between successive zero crossings. It corresponds to the appearance of a complex riding wave (a complicated cycle of an alternating signal). A vibration signal with a positive IF and envelope is considered to be a monocomponent signal: in each cycle, defined by the zero crossing, it involves only one mode of oscillation, and no complex riding waves are allowed (Huang *et al.*, 1998). A deterministic slow frequency modulated quasiharmonic, and a narrowband random vibration signal, are two typical examples of monocomponent signal. In this case the IF will not have the fast fluctuations induced by asymmetric waveforms.

Monocomponent signals belong to the group of narrowband signals with an IF always greater than zero. However, not every narrowband signal is a simple monocomponent signal. For example, a slow AM signal is a narrowband signal, but is not a monocomponent, because it can be decomposed into three simpler components (Section 4.5). All the examples of signals considered in Sections 4.5–4.8 can be decomposed into simpler components, each of which has no envelope or IF deviation, and a zero spectrum bandwidth.

5.2   Multicomponent signal

For more complicated vibration signals, the envelope and the IF vary rapidly in time and are not always positive. Often, these signals can be represented by the composition (sum, mixture, combination) of a small number of monocomponent signals (Boashash, 1992). Let us now consider the original signal $x(t)$, expressed as a sum of different monocomponents, each of which has a slow-varying instantaneous amplitude and frequency, so that

$$x(t) = \sum_1^k A_l(t) \cos \left(\int \omega_l(t) dt \right), \tag{5.1}$$

where $A_l(t)$ is the envelope (instantaneous amplitude) and $\omega_l(t)$ is the angular IF of the l component. In other words, the signal consists of k monocomponents, where each one has a constant or slowly varying amplitude $A_l(t)$ and an IF $\omega_l(t)$; and l indicates the number of components that have a different oscillatory frequency.

By construction, each monocomponent $A_l(t) \cos \left(\int \omega_l(t) dt \right)$ is an intrinsic mode of $x(t)$ with a simple oscillatory waveform described by the envelope $A_l(t)$ and the IF $\omega_l(t)$. If $l = 1$, the vibration is said to be a monocomponent signal; if, however, $l \geq 2$, then the vibration with a wideband spectrum has been referred to as a multicomponent signal.

For a multicomponent signal we propose the following definition: an asymptotic signal is referred to as a multicomponent composition if there exists even a single narrowband component such that its extraction from the composition decreases the average spectrum bandwidth of the remainder of the signal. This definition means that, after breaking a signal into its simplest components and taking out any component, the envelope and the IF of the residue part will have smaller deviations in time.

Having an initial multicomponent signal as a mesh of separate parts, the aim is to find a way to break it into the same parts with their individual envelopes and an IF. Obviously, any signal can be broken into an infinite number of constituents; however, it is desirable to express such a decomposition as the sum of proper simpler mono-components. Much time and effort was spent to design algorithms that decomposed the signal into elementary components (Hogan and Lakey, 2005). However, even obedience to almost all of the global conditions discussed earlier did not provide a scheme that showed how to obtain the decomposition given in (5.1). In the late 1990s, Huang (Huang *et al.*, 1998; Huang, Shen, and Long, 1999) introduced the Empirical Mode Decomposition (EMD) method for generating intrinsic modes that are almost monocomponents. The core of the EMD method is to identify the innate undulations belonging to different time scales and sift them out to obtain one intrinsic mode at a time. This can be achieved by using artificial extrema functions defined by a spline fitting of the local maxima and minima to discern waves that are riding on top of others (Huang, Shen and Long, 1999; Lai and Ye, 2003).

The problem of decomposing a nonstationary wideband vibration can be achieved by another technique, called the Hilbert Vibration Decomposition (HVD) method (Feldman, 2006). It is significant that the HVD method is based on a Hilbert transformation and does not involve any additional complicated signal-processing procedure. One of the most important results of the decomposition is an ability to preserve the signal phase content by constructing every initial component in the time domain and preserving all of its actual phase relations. The obtained combination of the simplest components with time and phase relations can provide us with some insight into a nonstationary vibration signal. Moreover, the method achieves an improvement in the analysis of a nonlinear dynamic system that cannot be accomplished by other methods. The individuality of the simplest components inside the vibration mixture helps us to choose the most effective decomposition method.

5.3 Types of multicomponent signal

The class of so-called almost periodic motions is a particular subclass of recurrent trajectories, and is of interest in nonlinear dynamics. A remarkable feature – which reveals the origin of these trajectories – is that each component of an almost periodic motion is an almost periodic function with well-studied analytical properties. An almost periodic function is uniquely defined "on average" by a trigonometric Fourier series $f(t) \approx \sum_{n=-\infty}^{\infty} a_n e^{i\lambda_n t}$, where λ_n are real numbers. If all λ_n are linear combinations (with integer coefficients) of a finite number of rationally independent elements from a basis of frequencies, then we have a particular case of almost periodic functions, namely *quasiperiodic* functions. A quasiperiodic signal, in this context, is a signal that consists of the sum of a given number of sinusoidal signals with known frequencies and unknown, time-varying, amplitudes and phases. This kind of quasiperiodicity emerges in vibration signals. In many practical situations, it is desirable that parameters of a quasiperiodic signal be estimated in real time. A continuous estimation of these parameters can be used, for example, for measuring, monitoring, or diagnostics purposes.

An arbitrary multicomponent signal as the sum of monocomponent signals can be conventionally divided into several types, depending on the behavior of the component's instantaneous characteristics:

Type I. Well-separated (non-crossing) and smoothed envelopes and IF trajectories. In this case the energy and frequency of every component are well concentrated (localized), and the components are not overlapping. These simple components resemble Fourier modes in the Fourier series, where every component can have a varying but non-crossing envelope and IF.

Type II. Well-separated, but rapidly changing envelopes and IF trajectories. In this case the instantaneous characteristics of components can jump in time or can be fast-varying functions. This type corresponds to a multicomponent signal composed of sequential segments with a step-varying envelope and IF.

Type III. Crossing envelopes and IF trajectories. In this case the envelope and/or the IF trajectories have single or multiple crossings between each other. From the standpoint of decomposition, this is the most complicated type of multicomponent signal. The decomposition result is not unique and is application dependent (Boashash, 1992).

5.4   Averaging envelope and instantaneous frequency

Along with the instantaneous characteristics, it is desirable to know the average value of the envelope and the IF over the analysis time. The most common average is an arithmetic mean value as a central tendency (middle, expected) measure of the data change over some interval $\bar{f} = (b - a)^{-1} \int_a^b f(t)dt$. Let us consider a mean value of the instantaneous characteristics of a multicomponent signal in the form of two harmonic compositions. In this case, the signal can be modeled as a weighted sum of the monocomponent signals, each with its own frequency and amplitude: that is, $X(t) = A_1 e^{i\omega_1 t} + A_2 e^{i\omega_2 t}$ with A_1, A_2, ω_1 and ω_2 being constant values. The envelope $A(t)$ and the IF $\omega(t)$ of the double-component vibration signal are (4.8), (4.9):

$$A(t) = \left[A_1^2 + A_2^2 + 2A_1 A_2 \cos(\omega_2 - \omega_1)t \right]^{1/2}$$

$$\omega(t) = \omega_1 + \frac{(\omega_2 - \omega_1)\left[A_2^2 + A_1 A_2 \cos(\omega_2 - \omega_1)t \right]}{A^2(t)}, \quad (A_1 > A_2; \omega_2 > \omega_1)$$

$$(5.2)$$

As mentioned, the composition envelope consists of two different parts – a constant part including the sum of the component amplitudes squared, and a fast-varying (oscillating) part. Integration of the envelope function (5.2) removes the fast-varying part and keeps the constant amplitudes $\overline{A(t)} = \left(A_1^2 + A_2^2 \right)^{1/2}$. So the average envelope is the square root of the arithmetic mean (average) of the component amplitudes squared of the signal components $\overline{A(t)} = \left(\sum A_i^2 \right)^{1/2}$.

The IF of the two tones also consists of two different parts: that is, the first component constant frequency ω_1 and a fast-varying asymmetrical oscillating part. However, the fast-varying asymmetrical oscillating part of the IF has an important feature. If we integrate the oscillating part with the integration limits corresponding to the full period of the cycle frequency $\left[0 \quad T = \frac{2\pi}{\omega_2 - \omega_1}\right]$, assuming that $A_1 > A_2$, we will obtain a definite integral equal to zero (Feldman, 2006)

$$\int\limits_0^T \frac{(\omega_2 - \omega_1)\left[A_2^2 + A_1 A_2 \cos(\omega_2 - \omega_1)t\right]}{A^2(t)} dt = 0. \tag{5.3}$$

The integral zero value means that the second rapid part disappears after averaging and has no influence on the average IF value (5.2). Because of this the remaining average value of the IF is only equal to the frequency of the first part, namely, the frequency of the largest harmonic ω_1: $\overline{\omega(t)} = \int_0^T \omega(t)dt = \omega_1 + 0$. Only in the particular hypothetical case of two identical components amplitudes will the average composition IF be equal to the average value of both component frequencies $\overline{\omega(t)} = 0.5\left[\omega_1(t) + \omega_2(t)\right]$ (Loughlin and Tacer, 1997). Wei and Bovik (1998) proposed an interesting study and interpretation of the average IF of a signal having no more than three dominant components well separated in the time domain. It was shown that the way the components are related and interact is crucial for the interpretation of the IF of a multicomponent signal.

In the case of the time-varying parameters of two quasiharmonics, the signal again can be modeled as the weighted sum of two monocomponent signals, each with its own slow-varying amplitude and frequency: $A_1(t)e^{i\int_0^t \omega_1(t)dt} + A_2(t)e^{i\int_0^t \omega_2(t)dt}$. For averaging the fast-varying asymmetrical oscillating part of the IF (5.3) we replace the integration over the full cycle period by a convolution with a lowpass filter. The lowpass filtering – instead of the integration – will cut down the fast asymmetrical oscillations and leave only a slow-varying frequency of the main signal component $\omega_1(t)$. Small fast output oscillations after the filter can always be neglected in comparison with the main slow member of the IF. Thus, the averaging of the IF by a lowpass filtering makes it possible to detect the frequency of the largest energy component. The same is also true for three and more components in a composition. In the general case, the IF will have a more complicated form (Nho and Loughlin, 1999; Suzuki et al., 2006), but, again, the averaging or the lowpass filtering will extract only the IF of the single largest energy component.

This interesting property of the IF offers the simplest way to estimate the frequency of the largest vibration component. An averaging, or smoothing, or lowpass filtering of the IF of the vibration composition will cut down asymmetrical oscillations and leave only the slow-varying frequency of the main vibration component. This result is a central condition allowing the development of a new effective Hilbert Vibration Decomposition method (Feldman, 2006).

5.5 Smoothing and approximation of the instantaneous frequency

The IF of the multicomponent signal is a complicated function of time; many attempts were made to smooth the IF by creating approximations that would capture important patterns in the data, while leaving out incidental peaks and noise. The simplest of the different algorithms for the IF smoothing and approximation is the Teager–Kaiser energy operator $\Psi(x) = \dot{x}^2 - x\ddot{x}$ (Teager, 1980). The energy operator $\Psi(x)$ derives the signal temporary frequency according to a simple three-point algorithm $\omega = \left[\Psi(\dot{x})/\Psi(x)\right]^{1/2}$. The Teager–Kaiser energy operator was developed by Teager during his work on speech AM–FM modeling and was first introduced systematically by Kaiser (1990). The three-point algorithm is able to recover the energy of a sampled signal and – in certain conditions – the IF. Girolami and Vakman (2002) proposed another three-point method, which gives excellent results for the estimation of both the IF and the envelope of a signal that can be considered as semi-local, because filtering defines its integrality. However, filtering affects only short intervals around a given instant, giving the method its local character. In theory, the method is equivalent to the analytic signal and, in practice, gives a very good approximation of the results one can obtain with the analytic signal. Being quasilocal, the method eliminates the doubts associated with the global nature of the analytic signal.

When the energy operator is applied to signals produced by a simple harmonic mass-spring oscillator, it can track the oscillator's energy (per half unit mass), which is equal to the squared product of the oscillation amplitude and frequency. At each instant the energy operator estimates the central frequency and amplitude of the signal by using the output values of the energy operator applied to the signal $x(t)$ and the signal derivative $\dot{x}(t)$. The energy operator was compared to the HT tracking modulations in speech signals as an alternate approach to the demodulation signal amplitude and frequency (Potamianos and Maragos, 1994). The AM–FM modulation model, and the energy separation algorithm, have been used successfully to determine the center values of the formant frequencies in a speech segment. It was shown that it had many attractive features such as simplicity, efficiency, and adaptability to instantaneous signal variations; the energy operator approach has a smaller computational complexity and a faster adaptation due to its instantaneous nature.

The energy operator can approximately estimate the squared product of the amplitude and frequency signals, but only if the signal's instantaneous parameters do not vary too fast or too greatly with time compared to the carrier frequency. This means that the energy operator is not suitable for the case of a multicomponent vibration signal with a high rate of IF variation when all existed components will affect the central frequency. Before applying the energy operator to the demodulation of the multicomponent resonance signal, one needs first to isolate the signal through a narrow bandpass filter (Potamianos and Maragos, 1994). For a single harmonic, or narrowband signal, while the IF is a positive function and the signal has the same numbers of zero crossings and extrema, the energy operator works well and its estimations coincide with the IF. But it has some drawbacks; it cannot work correctly on a multicomponent signal (Vakman, 1996).

Another modern algorithm is the Normalized HT developed recently by Huang et al.(2009). The algorithm incorporates the EMD decomposition and normalization method to reduce the signal to the Intrinsic Mode Functions. Each decomposed Intrinsic Mode Function $x_l(t)$ has its own envelope $e_l(t)$. The normalization procedure includes receiving normalized fast-oscillating quasiharmonic data $\cos\psi_1(t)$ through the following repeated steps:

$$\cos \psi_1(t) = \frac{x(t)}{e_1(t)}; \quad \cos \psi_2(t) = \frac{\cos \psi_1(t)}{e_2(t)} = \frac{x(t)}{e_1(t)e_2(t)}; \dots$$

$$\cos \psi_n(t) = \frac{x(t)}{e_1(t)e_2(t)\dots e_n(t)}.$$

The last step results in a normalized instantaneous phase $\cos \psi_n(t) = \cos \psi(t)$ as the empirical frequency modulated part of the initial signal $x(t) = A(t)\cos \psi(t) = A(t)e_1(t)e_2(t)\dots e_n(t)$. The normalized IF $\dot{\psi}(t)$ computed from the normalized data will be the empirical frequency of the modulated signal. The empirical IF is smoother and does not include fast fluctuations and other overshoots. Similar smoothing effect can be achieved with signal filtering procedures such as the Savitzky–Golay algorithm (Savitzky and Golay, 1964).

In many engineering cases the experimentally obtained vibration data inevitably contains random noise and other instrumental errors. Owing to the signal derivative, even a minor noise can easily amplify the inaccuracy of the envelope – especially the IF estimation – and deteriorate the quality of data. To extract smooth and meaningful information, a curve-fitting technique can be used in such engineering applications. An example of the envelope curve-fitting and de-noising technique, based on local-maxima interpolation, is given in Luo, Fang, and Ertas (2009). The instantaneous characteristics curve fitting, after the direct HT calculation, can greatly improve the envelope estimation by removing severe distortions and errors.

Signal smoothing and filtering techniques are especially attractive in a vibration analysis. In addition to the empiric, or artificial, algorithms we can suggest another simple and natural way to estimate smooth bounds of the instantaneous characteristics.

5.6   Congruent envelope

The aim of this section is to provide a new approach to the estimation of the acting smooth upper and lower bounds of a vibration composition. A vibrating signal regularly changes its direction, and the specific is not only the signal parameter alternation, but also the incessant alternation of each instantaneous function. In reality, both the envelope and the IF of the composed signal also oscillate and periodically touch their extremum value – even more frequently than the signal. The idea is to consider each instantaneous function as a new oscillation variable having its own new *envelope of the envelope* (EOE) and a new *envelope of the IF* (Feldman, 2007a). These new smooth congruent envelope functions are of interest for multicomponent vibration-compounding ultraharmonics (superharmonics) – that is, the high-frequency components of frequencies that are integer multiples of the largest energy frequency component.

To explain the meaning of the EOE we will use the concept of the *phase congruency* (Kovesi, 1999). The concept is similar to coherence, except that it applies to functions of different frequencies. The phase congruency describes points of local symmetry and asymmetry in the multicomponent signal – according to the special arrangements of the phase that manifest at these points. Multiple frequencies included in a signal composition can be uncovered by a direct spectral analysis, for instance. It was noted that waveform features had many of their frequency components in or out of the same phase. For example, all high-frequency superharmonics decomposed from the square wave are phase congruent and have the same $0°$ or $180°$ phase shift. Shifting any of the cosine components of the waveform by an angle that is neither zero nor a multiple of $360°$ will result in a severe waveform distortion. It is clear that the EOE requires an estimation of the phase relations between different frequencies, but in the same local points. These local points are extrema points of the largest energy component. Different frequencies can be compared by determining each of the integer multiple superharmonics of the largest energy component. Therefore, geometrically, the EOF means the tangent or the upper bound of the composition at the local points where the largest energy component has an extrema. Therefore, the EOE function can also be called a *congruent envelope* (Feldman, 2011).

For an estimation of the EOE level of a multicomponent signal we will also use the triangle inequality theorem (Dragomir, 2005). Let us consider the sum of two arbitrary phasors. The triangle inequality determines the best upper estimation of the sum of two functions in terms of individual functions: $|X_1(t) + X_2(t)| \leq |X_1(t)| + |X_2(t)|$. The right-hand part of the triangle inequality states that the sum of the moduli (the envelopes) of any two phasors is greater than the modulus (the envelope) of their sum. This means that the upper amplitude bound of the composition is equal to the sum of the amplitude components. There is also a lower envelope estimation that can be determined using the inverse triangle inequality, which states that for any two functions $|X_1(t) - X_2(t)| \geq |X_1(t)| - |X_2(t)|$. Consequently, the lower amplitude bound of the composition is equal to the difference of the amplitude components. The new EOE function as a tangent to the envelope forms the upper bound of the largest energy component.

For two pure harmonics the envelope of the composition is a slowly time-varying periodic function (5.2) $A(t) = \left[A_1^2 + A_2^2 + 2A_1 A_2 \cos(\omega_2 - \omega_1)t \right]^{1/2}$. The envelope maxima values are equal to the algebraic sum of amplitudes, so the EOE will be the horizontal straight line that touches the points of a maximum. For two quasiharmonics, the EOE of the composition will vary slowly in time according to the slow variations of the amplitudes of two components $A_{EOE}(t) = A_1(t) \pm A_2(t)$.

As an example, we will estimate the EOE in the case of a composition of two rotating phasors with different frequencies and phase shifts. Let a vibration signal be composed of two harmonics, each with a slow variable amplitude and a constant frequency (Figure 5.1a): $x(t) = A_1(t) \sin \omega_1 t - A_2(t) \sin 3\omega_1 t$. According to (5.2) the acting extrema points of the envelope corresponding to $\omega_1 t = \pi/2$ produce a straight line, and the resultant EOE also will take the form of a straight line, $A_{EoE}(t) = A_1(t) + A_2(t)$. The results were achieved by a signal decomposition and an estimation of the EOE through the direct summation of all the envelopes. If a vibration signal is composed of two other harmonics: $A_1(t) \sin \omega_1 t + A_2(t) \sin 3\omega_1 t$, the resultant EOE will also take the form of a straight line (Figure 5.1b). But now

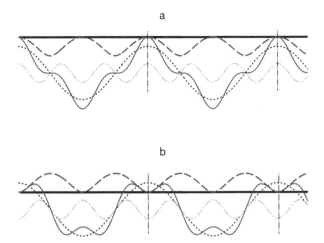

*Figure 5.1   The sum of two in-phase (a) and out-of-phase (b) harmonics: the enve-
lope (- -), the congruent envelope (——), the largest (. . .), the secondary (-··-) harmonic
(Feldman, ©2011 by Elsevier)*

the acting extrema points of the envelope produce a lower straight line $A_{\text{EoE}}(t) = A_1(t) - A_2(t)$ corresponding to $\omega_1 t = \pi/2$.

Let $x(t)$ be a general l-component real signal with real amplitudes and instantaneous frequencies (5.1). Naturally, this multicomponent phasor as a composition $X(t) = \sum_{l=1}^{N} A_l(t)e^{j\phi_l(t)}e^{j\int\omega_l(t)dt}$ will have a more complicated and faster varying envelope function. Here we have a signal in analytic form and a complex amplitude (envelope) for every component $A_l(t)e^{j\phi_l(t)}$. For a signal formed from the summation of many components, we can invoke the generalized triangle inequality property for complex functions. The triangle inequality states that the sum of the moduli of complex numbers is greater than the modulus of the sum of these complex numbers: $\left|\sum_{l=1}^{N} C_l\right| \leq \sum_{l=1}^{N} |C_l|$. In our case, the modulus on the left-hand side is the signal envelope $A(t)$ of the composition $X(t)$; the sum of the moduli on the right-hand side is the algebraic sum of the component envelopes, thus yielding: $A(t) \leq \sum_{l=1}^{N} A_l(t)\cos\phi_l(t)$. The EOE aggregates two or more component envelopes taken with regard to their phase functions $\cos\phi_l(t)$:

$$A_{\text{EOE}}(t) = \sum_{l=1}^{N} A_l(t)\cos\phi_l(t), \qquad (5.4)$$

where $A_l(t)$ and $\phi_l(t)$ are the envelope and the phase of the l component, respectively. The envelope of the sum of signals is equal to the sum of the envelopes if all the components are congruently in phase with the largest component.

The new EOE defined as a tangent curve to the local extrema touches only the maximum points of the largest component during every period. Such a double envelope represents the acting maximum height (intensity) of the largest signal component and forms a shape of the time variation of the acting extreme points. In other words, the EOE curve indicates the extrema (maximum) points of the l-factor composition, obtained by the algebraic summation of all the intrinsic components. The EOE varies with time much more slowly than the signal envelope itself. To estimate the EOE, we need first to produce a decomposition of the signal and then to construct an algebraic summation of the envelopes of all the decomposed components. Such a successive signal decomposition (disassembling), and the subsequent summation (re-assembling) of the component envelopes, generates the desirable EOE as a slow function of time. Some examples of the congruent approach are given in Section 6.8.5 and Chapter 11.

5.7   Congruent instantaneous frequency

The IF of the multicomponent signal is a fast-varying and complicated function of time. For simplicity let us consider the IF of the composition of two pure harmonics as a time-varying periodic function (5.2): $\omega(t) = \omega_1 + \frac{(\omega_2 - \omega_1)\left[A_2^2 + A_1 A_2 \cos(\omega_2 - \omega_1)t\right]}{A^2(t)}$. The corresponding IF extrema values are equal to $\omega_1 + (\omega_2 - \omega_1)\left(A_2^2 \pm A_1 A_2\right)/A_{\text{EOE}}^2 = \omega_1 A_1/A_{\text{EOE}} \pm \omega_2 A_2/A_{\text{EOE}}$. It is clear that the IF extrema values are a combination of all existing frequency components. The IF of the composition basically depends on the largest energy component, and therefore the extrema of the IF coincide with the extrema points of the largest energy component. Let us build a new function in the form of an envelope as a tangent to the varying IF in the local extrema points of the largest energy component. We will call this new function the *envelope of the IF,* or congruent IF, because it shapes the upper or lower bounds of the IF in the local extrema points of the largest energy component (Figure 5.2). Thus, this new congruent IF can be expressed as a mixture of signal components of different

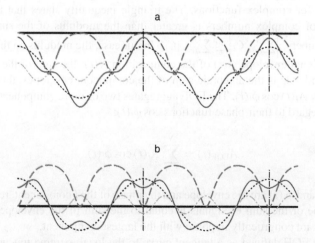

Figure 5.2   The sum of two in-phase (a) and out-of-phase (b) harmonics: the IF (- -), the congruent frequency (——), the largest (. . .), the secondary (- · -) harmonic (Feldman, ©2011 by Elsevier)

frequency. For only two components the congruent IF is the algebraic sum of two weighted component frequencies $\omega_{EOIF} = \omega_1 A_1 / A_{EOE} \pm \omega_2 A_2 / A_{EOE}$, where ω_{EOIF} is the envelope of the IF, A_{EOE} is the EOE, A_1, A_1 are the harmonic amplitudes, and ω_1, ω_2 are the harmonic frequencies.

A relative weight of each frequency component to the concluding congruent IF depends on its amplitude ratio, frequency, and phase relation. Thus, in a general case of many vibration components, the congruent IF aggregates them all with regard to their phase functions $\cos \phi_l(t)$:

$$\omega_{EOIF}(t) = \sum_{1}^{N} \left[\omega_l(t) A_l(t) \cos \psi_l(t) / A_{EOE}(t) \right], \qquad (5.5)$$

where ω_{EOIF} is the congruent IF, $A_l(t)$ and $\omega_l(t)$ are the l component envelope and frequency, respectively. The phase angle $\phi_l(t)$ is the angle between the first largest and the l component. Again, to consider the cosine components at multiply-spaced frequencies and their phase shifts in relation to each other, we will use the concept of phase congruency (Kovesi, 1999). In the previous example of two components, the congruent IF was the horizontal straight line that touched the extrema points of the IF (Figure 5.2). For a nonstationary composition of quasiharmonics the congruent IF will vary slowly in time according to the slow variations of the amplitudes and frequencies of the components.

The proposed congruent functions can be simply straight or slow-varying bound lines even for very rich vibration contents. In the cases of a triangle (Figure 5.3) or square (Figure 5.4), the waves show a series of harmonics (depicted with filled tints) which, when summed together, form a triangle or square wave. For the triangle and square waveforms, we noted a dissimilar relationship between the phase congruency

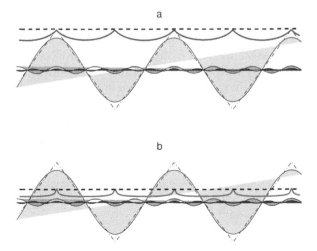

Figure 5.3 *The triangle signal and its high harmonics: the envelope (a, ▬) and the congruent envelope (a, - -); the IF (b, ▬) and the congruent IF (b, - -) (Feldman, ©2011 by Elsevier)*

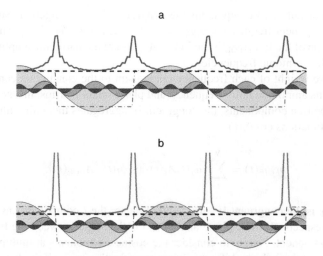

Figure 5.4   The square signal and its high harmonics: the envelope (a, ___); the congruent envelope (a, --); the IF (b, ___) and the congruent IF (b, --) (Feldman, ©2011 by Elsevier)

of the localized frequencies. Thus, the entire integer multiple ultraharmonics are in-phase congruent for a triangle wave (Figure 5.3); whereas the square wave ultraharmonics phase values alternate from in phase to out of phase (Figure 5.4). The phase congruent frequency components – when aggregated – give a smooth EOE function. It is interesting that the EOF curve of the square wave is mapped only by a single point bounding at the bottom on the envelope vs. IF plot (Figure 6.3).

In this section we described the geometrical meaning of both smooth functions of the vibration signal: the congruent envelope and the congruent IF. This result represents the congruent envelope and the IF functions of the multicomponent signal in terms of its components. The next section is devoted to a vibration system analysis and will show the physical meaning of these congruent envelopes; it is significant for nonlinear vibration systems. The congruent functions play an important role in the asymmetric signal analysis (Section 11.7) and in the identification of nonlinear vibrations when considering their high-frequency nonlinear components (Chapter 11). The HT decomposition procedures splits the signal into a number of components with multiple signatures. Development of the congruent characteristics, as the opposite tendency, aggregates these multiple signatures and expresses them as smooth combined EOE functions having a clear geometrical and physical meaning. It seems that with these results some of the problems related to the presentation of the instantaneous characteristics are resolved, and a simple and natural congruent notation is thus taking another step toward its maturity.

5.8   Conclusions

Often, vibration signals can be represented by a composition (sum) of a small number of its monocomponent signals. Therefore, in engineering and research work it is necessary to analyze multipart vibration concurrently composed from large pieces of elementary vibration signals. These elementary components – as a number of varying harmonics – might appear, for example, as modulated sidebands around individual carrying frequencies in the frequency domain. The more harmonics, the more intensive is the modulating effect and distortion of the observed compound signal.

For a multicomponent signal we propose the following definition: an asymptotic signal is referred to as a multicomponent composition if there exists even a single narrowband component, such that its extraction from the composition decreases the average spectrum bandwidth of the remainder of the signal.

When an initial multicomponent signal is a mesh of separate parts, the aim is to find a way to break it into the same parts with their individual envelopes and IF. It is desirable to know the average value of the envelope and the IF over the analysis time, as well as the instantaneous characteristics. A HT method for simultaneously examining the signal components of vibration is described in the following chapter.

6

Local and global vibration decompositions

6.1 Empirical mode decomposition

Before analyzing a multicomponent vibration, the first common step is to decompose, or separate, the original signal into its elementary and fundamental constituents for performing further signal-processing operations separately on each component. Classic examples of a decomposition are a signal filtering with a high/lowpass or a narrowband filter, a spectral analysis that breaks signals into harmonic components, and a wavelet (time-scale) decomposition that projects the signal on the set of wavelet basis vectors. Many real-world oscillating signals are nonstationary, which makes common basic decomposition techniques, such as the Fourier decomposition or the wavelet decomposition, unsatisfactory since the basic functions of the decompositions are fixed and do not necessarily match the varying nature of the signals. In the case of a complicated time-varying waveform, the main purpose of a signal decomposition is not to represent a true frequency distribution, but rather to represent complicated waveform contents in the time domain.

Recently, Huang *et al.* (1998) proposed the EMD method to extract monocomponents and symmetric components, known as IMF (Intrinsic Mode Functions), from nonlinear and nonstationary signals. The term "empirical," chosen by the authors, emphasizes the empirical essence of the proposed identification of the IMF by their characteristic time scales in the complicated data. Application of the EMD method is now referred to as a Huang–Hilbert Transform (HHT) in the technical literature (Daetig and Schlurmann, 2004; Attoh-Okine *et al.*, 2008).

The EMD is a new adaptive technique representing nonlinear and nonstationary signals as sums of simpler components with amplitude and frequency modulated parameters. This technique is also capable of displaying the overlap in both time and frequency components that cannot be separated by other standard filtering techniques

Hilbert Transform Applications in Mechanical Vibration, First Edition. Michael Feldman.
© 2011 John Wiley & Sons, Ltd. Published 2011 by John Wiley & Sons, Ltd.

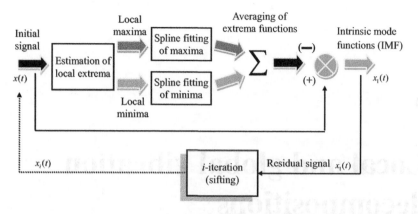

Figure 6.1   Block diagram of the EMD method

or traditional Fourier methods. The method is able to visualize a signal energy spread between available frequencies locally in time, thus resembling the wavelet transform. Hence, it was immediately applied in diverse areas of signal processing. In the field of sound and vibration of mechanical systems, the EMD method has been also applied widely for diagnostics and structural health monitoring, as well as in analysis and identification of nonlinear vibration, mainly for rotating systems with typical elements such as bearings and gears.

The EMD method (Huang *et al.*, 1998) automatically generates a collection of intrinsic mode functions that satisfy two conditions: (a) in a complete data set, the number of extrema and the number of zero crossings must be equal or not differ by more than one; and (b) at any point, the mean value of the envelope defined by the local maxima, and the envelope defined by the local minima, is zero. The EMD algorithm requires the following procedures at every iteration step (Figure 6.1): (1) estimation of all local extrema; (2) spline fitting of all local minima and maxima, ending up with two (top and bottom) extrema functions; (3) computation of the average function between maxima and minima; (4) extraction of the average from the initial signal; and (5) iteration on the residual (the sifting procedure). When applied to a nonstationary signal mixture, the EMD algorithm yields efficient estimates for individual components with slow varying instantaneous amplitude and frequency of each component.

For illustration purposes, let the signal be a composition of two harmonics, each one with its own amplitude and frequency (Figure 6.2a). The first subplot shows the signal composition, both the upper and the lower envelopes which are out of phase, and local top and bottom extrema points. It is easy to see that the local top and bottom extrema points specified by the triangle up and down signs, respectively, are not located on the envelope functions.

The spline-fitted local minima and maxima, connecting the top and bottom extrema functions, are shown in Figure 6.2b. Their averaging extracts the low-frequency component; the residual part of the composition will be the second fast-frequency harmonic (Figure 6.2c). The EMD method successfully decomposes the amplitude and frequency of each intrinsic function from the initial composition of far

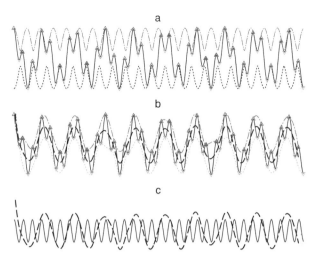

Figure 6.2   The sum of two harmonics: the initial signal (a, —), the upper (···) and lower (···) envelope, the top (△) and bottom extrema points (▽); the top (b, —) and bottom (b, ···) extrema, the mean value between the top and bottom extrema (- -); the EMD decomposed first harmonic (c, - -), the second harmonic (—) (Feldman, ©2011 by Elsevier)

frequency components, but it cannot decompose close frequency components (Rilling and Flandrin, 2008).

Each envelope and IF of every decomposed IMF can be considered as a pair of time functions. The obtained envelopes and IF of all decomposed components joined together form the frequency–time distribution of the amplitude known as the Hilbert spectrum $H(\omega, t)$ (Huang *et al.*, 1998). The Hilbert spectrum can be plotted as two separated graphs of amplitude and frequency ensembles varying in time. This way we achieve the best possible time and frequency resolution of analysis. Both time-varying amplitudes and frequencies can be plotted in combination as a single 3D plot presenting the contribution of each component to the signal composition.

We can also exclude the time, and present the Hilbert spectrum as a continuous amplitude–frequency plot. Sometimes it is suggested that each decomposed IF should be divided into discrete bins and then plot the bins all together versus their envelope values. Such a frequency–amplitude distribution will look very similar to the classic Fourier spectrum. However, such a spectrum-like presentation will lose the frequency resolution, not to mention the time dependency of every component. The amplitude-integrated distribution of all pairs $\int H(\omega, t)dt$ is known as the *Hilbert marginal spectrum* (Huang *et al.*, 1998). The Hilbert spectrum offers a measure of amplitude contribution from each frequency and time, while the marginal spectrum offers a measure of the total amplitude contribution from each frequency (Yu, Cheng, and Yang, 2005).

In many cases of vibration analysis, the Hilbert spectra is much simpler and clearer than the Envelope vs. IF plot. For example, consider the square wave signal in Figure 5.4. Its Envelope vs. IF graph (shown in Figure 6.3) has a rather complicated and indistinct shape. The Hilbert spectrum of the same square wave (Figure 6.4)

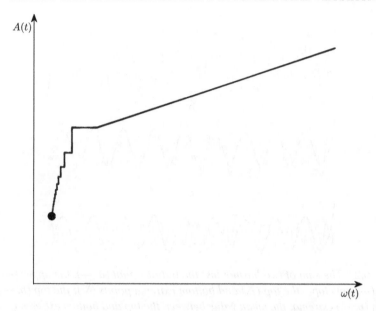

Figure 6.3 The square wave envelope vs. the IF (—); the congruent envelope vs. the congruent IF (•) (Feldman, ©2011 by Elsevier)

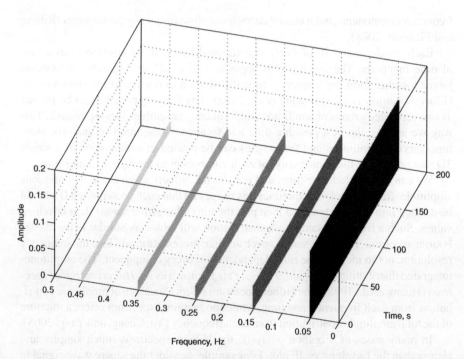

Figure 6.4 The square wave Hilbert spectrum (Feldman, ©2011 by Elsevier)

enables us to represent the amplitudes and instantaneous frequencies as functions of time in a three-dimensional plot, where every envelope and the IF are simply straight lines. The time–frequency distribution of instantaneous amplitudes and frequencies, designated as the Hilbert spectrum, requires a preliminary decomposition of the initial signal.

In addition to successful applications of the EMD methods, a lot of attempts have been made to improve, or at least to modify, the original method. Some of these attempts suggested replacing the original cubic spline fitting by other kinds of interpolation or by a parabolic partial differential equation. Other researchers proposed applying correlation functions, FFT spectrum, Teager's energy operators, or other regularizations for detecting amplitude and frequency peaks. Still others suggested using the iterated Hilbert transform method that deals only with AM–FM modulated multicomponent signals, or the Local Mean Decomposition iterative approach that uses a standard moving averaging to successively extract smoothed components. All these efforts to modify the EMD were made, and the results were published even before the method was explained theoretically or substantiated analytically.

In parallel, sophisticated studies devoted to analyzing the essential shortcomings of the EMD and its restrictions – in comparison with other decomposition methods – began to appear. One of the first limitations of the method was a rather low-frequency resolution (Wu and Huang, 2004) – the EMD can resolve only distant spectral components differing by more than an octave. Another weak point of the method was that its application produced false artificial components not presented in the initial composition. However, the newest Ensemble Empirical Mode Decomposition (EEMD) method (Wu and Huang, 2009) largely overcomes the false mode mixing problem of the original EMD and provides physically unique decompositions.Recent progress in the development of EMD may make it an important tool in the decomposition of nonstationary vibration – to be considered before applying the HT or any other signal-processing technique.[1] In the last few years, there are several major achievements in the EMD method development summarized in (Huang and Wu, 2008), the most important ones among them being the characteristics of EMD and the statistical significance test of IMFs (Wu and Huang, 2004), the ensemble EMD (EEMD) (Wu and Huang, 2009), and the EMD-based normalized instantaneous frequency calculation (Huang *et al.*, 2009).

6.2   Analytical basics of the EMD

All earlier publications agree that the EMD is defined empirically only – by its algorithm. At present, it does not propose an analytical formulation that would allow theoretical analysis and analytical performance evaluation. A typical case of a composition of a harmonic and a slow-varying aperiodic trend (like DC) – which does allow a simple analytical foundation – is an exception. The HT projection of such a composition looks even simpler than the initial signal because the HT of the constant is equal to zero. Therefore, it is easy to show analytically how the EMD removed the

---

[1] The EMD/EEMD programs are published on the Internet as the MATLAB® code files (Wu, 2010)

slow trend from the composition. A recent study by Kizhner *et al.* (2006) offers a theoretical explanation of the EMDs behavior. The authors attempt to build the basic theory of decomposition of artificially created fast and slow-varying components. Despite its considerable number of hypotheses and proofs, this work does not quite convincingly explain why the fastest component is sifted out first in the EMD. During the last decade, serious mathematic works, (Wu and Huang, 2004, 2009; Sharpley and Vatchev, 2006; Kizhner *et al.*, 2006; Flandrin, Rilling, and Goncalves, 2004; Rilling and Flandrin, 2008), have been dedicated to detailed analysis of the local EMD method. However, a simple but important theoretical question remains: why is spline fitting of local extrema able to generate the simplest components?

6.2.1   Decomposition of a harmonic plus DC offset

To understand why the spline fitting of local extrema is able to extract the simplest components, we will consider two main cases of the signal composition: a harmonic plus the DC offset and the sum of two different harmonics. In the case of the DC offset $x(t) = a(t) + A_1(t) \cos \omega t$, the spline fitting of maxima $\omega t = 0 \pm 2\pi n$ produces a straight line $A_{max} = A_1(t) + a(t)$ and interpolation between minima $\omega t = \pi \pm 2\pi n$ produces another straight line $A_{min} = -A_1(t) + a(t)$. The mean value of these two lines will generate the initial offset function $0.5 (A_{max} + A_{min}) = a(t)$. Extracting the obtained offset from the initial composition yields the initial harmonic $x(t) - a(t) = A_1(t) \cos \omega t$. In a similar way, the EMD method used on each iteration successfully separates the harmonic and the slow aperiodic component.

6.2.2   Decomposition of two harmonics

The next example will explain theoretically and in a simple way why the EMD operates for harmonic functions and why it selects out the highest frequency oscillation, leaving the lower frequency oscillation in the signal. It is already known that the EMD has a poor frequency resolution and does not allow separation of the closest harmonics due to the existence of a critical frequency limit. These questions are considered in depth in the work of Rilling and Flandrin (2008); they examine the case of decomposing two harmonics. The work provides theoretical and experimental proof that there exist three domains of amplitude–frequency harmonics relations: (1) the components are well separated and correctly identified; (2) the harmonics are considered as a single waveform; and (3) the EMD does something else. However, this work is theoretically based on rather complicated Fourier transform models.

The notion of two fast and slow harmonics appears to be rather fruitful. A most realistic vibration or sound signal can be seen as consisting of a linear combination of two or more sinusoids. In the present study we will approach the theoretical foundation of the EMD decomposition in the simplest way – through direct analysis of the harmonics separation. For the sum of two harmonics, we will show why the low-frequency harmonic remains while the high-frequency harmonic is sifted out first in the EMD procedure. We will also find a theoretical limiting frequency resolution using the method.

Every maximum (top extremum) and minimum (bottom extremum) point of the function, known collectively as the set of local extrema points x_{extr}, is uniquely defined by the first derivative when the slope of the tangent is equal to zero. The initial signal and its envelope have common tangents at points of contact, but the signal never crosses the envelope. The common points of the contact between the signal and its envelope do not always correspond to the local extrema of a multicomponent signal. The local extrema always have a zero tangent slope, but the common points of the contact can have a nonzero value for the tangent slope. The distance between the common points of contact and the extrema points of the signal plays a dominant role in explaining the EMD mechanism. In effect, without such a difference between the envelope and the local extrema, the sum of maxima and minima curves required by the EMD would be equal to zero, just like the zero sum of the upper and the lower envelopes. A combination of two harmonics is a rather representative case of a signal composition. This case provides us with means to discover and prove some important features of the EMD.

6.2.3   Distance between envelope and extrema

Let consider the case of a signal composition as a sum of two harmonics: $x(t) = A_1 \cos \omega_1 t + A_2 \cos \omega_2 t$, where A_1, A_2 are amplitudes and ω_1, ω_2 are frequencies of the harmonics. The vertical position of the local extrema according to (4.17) depends on the variable $\frac{A(t)\omega(t)}{\dot{A}(t)}$, which in the case of two harmonics has the same period as in a variation of the envelope $2\pi/(\omega_2 - \omega_1)$: $\frac{A(t)\omega(t)}{\dot{A}(t)} = -\frac{\omega_1 A_1/A_2 + \omega_2 A_2/A_1}{(\omega_2 - \omega_1)\sin(\omega_2 t - \omega_1 t)}$
$-\frac{\omega_2 + \omega_1}{\omega_2 - \omega_1} \cot(\omega_2 t - \omega_1 t)$. Multiplying the envelope and the cosine projection generates the position as a periodic function with the same period $2\pi/(\omega_2 - \omega_1)$:

$$x_{extr}(t) = A(t)\frac{A(t)\omega(t)}{\dot{A}(t)\sqrt{1 + \frac{A^2(t)\omega^2(t)}{\dot{A}^2(t)}}}. \tag{6.1}$$

This vertical position of the local maxima varies from its highest to its lowest position during a single period, thus specifying a band with all possible local extrema.

When the instantaneous frequency of the composition becomes negative $\frac{A_2}{A_1} > \frac{\omega_1}{\omega_2}$, the continuous vertical position is a monotonic function with a top maximum always equal to the sum of amplitudes of both harmonics: $x_{max(top)}(0) = A_1 + A_2$ and with a bottom maximum position that is equal to the difference of amplitudes but has a negative value (Figure 6.5a): $x_{max(bottom)} = -A_1 + A_2$.

For an IF that is always positive when $\frac{A_2}{A_1} < \frac{\omega_1}{\omega_2}$, the top vertical position of maxima is again equal to the sum of the amplitudes: $x_{max(top)} = A_1 + A_2$. But now the vertical position decreases monotonically only until the intermediate bottom position $x_{max(bottom)i} = (A_1^2 + A_2^2 - A_1^2\omega_1^2/\omega_2^2 - A_2^2\omega_2^2/\omega_1^2)^{\frac{1}{2}}$ (Figure 6.5b). During the remainder of the period, the resultant vertical position jumps down to a negative symmetric value $-x_{max(bottom)i}$, specifying a second isolated band with a negative local maxima. A vertical position for the second band at the end of the period monotonically continues to the negative extreme bottom position $x_{max(bottom)} = -A_1 + A_2$ (Figure 6.5b). Theoretically, the values of the extrema deviation from the envelope

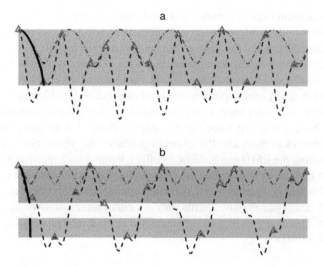

Figure 6.5   The vertical position of the local maxima (—), the initial signal (--), the upper envelope (···) and the top maxima (△): the negative IF (a) ($A_1 = 1$, $\omega_1 = 1$, $A_2 = 0.6$, $\omega_2 = 1.8$), the positive IF (b) ($A_1 = 1$, $\omega_1 = 1$, $A_2 = 0.25$, $\omega_2 = 3.9$)

depend on two ratios: the envelope, and the frequency of the harmonics. For a very small amplitude of the second harmonic $A_2 \leq 0.3 A_1 \omega_1 / \omega_2$, this intermediate bottom position does not differ much from the smallest envelope value $A_1 - A_2$. This means that the positive maxima points of the first band will always lie on the envelope. For other ratio range of the harmonics parameters $0.3 A_1 \omega_1 / \omega_2 \leq A_2 < A_1 \omega_1 / \omega_2$ the local maxima will wander vertically more and more from the envelope.

6.2.4   Mean value between the local maxima and minima curves

In general, the initial signal composition, as the sum of two harmonics, can be written as

$$x(t) = A_1 \cos \omega_1 t + A_2 \cos (\omega_2 t + \varphi) \tag{6.2}$$

where φ is the initial phase shift angle. The first derivative of the signal is

$$\dot{x}(t) = -A_1 \omega_1 \sin \omega_1 t - A_2 \omega_2 \sin (\omega_2 t + \varphi) \tag{6.3}$$

Every zero-crossing of the first derivative corresponds to the existence of a local extremum of the initial function. At a certain moment t_i, when the first derivative is equal to zero, a single local extremum maximum $x_{\max}(t_i)$ occurs. The closest single local minimum $x_{\min}(t_j)$ occurs at another certain moment t_j, so each closest maximum or minimum exists at different moments ($t_i \neq t_j$). However, the EMD method requires that both the top and bottom extremum curves be constructed of synchronous moments. For every top maximum we need to construct its virtual synchronous bottom pair, and, correspondingly, for every bottom minimum, its virtual synchronous top pair. As we know, the original EMD method builds such a synchronous top and bottom line by using cubic spline interpolation.

Let us analyze two closest neighbor extrema. For simplicity, let us find the closest minimum point from the left and right of each maximum. Then we estimate the mean (median) value of these two neighboring minimum points before and after the maximum, thus yielding the desired virtual synchronous bottom pair for each maximum. By connecting all virtual synchronous pairs, we obtain the bottom extremum line required by the EMD. The proposed simplest short straight line length fitting makes it possible to analyze and understand the main properties of the EMD.

By analogy, the mean value of two neighboring maximum points before and after the minimum will produce the desired virtual synchronous top pair of the minimum. As a result, for the initial signal composition with two sets of maxima and minima we will get two corresponding synchronous top and bottom lines constructed from short straight line lengths (Feldman, 2009a).

Originally, the EMD required the arithmetic mean value of the top and bottom lines to be computed. As shown, the extrema wander throughout the signal values, so the mean value will depend on the current position of the local extrema (6.1). Let us describe two different extreme cases: (a) the highest current position of the local maximum when $t_i = 0$, $\varphi = 0$, and (b) the lowest current position of the local maximum when $t_i = \pi$, $\varphi = \pi$. All other positions of the local maximum between these two extreme cases will exhibit only intermediate behavior.

The case of $t_i = 0$, $\varphi = 0$

In this case the initial maximum value is $x_{\max}(0) = A_1 + A_2$ and the initial value of the first derivative is $\dot{x}(0) = 0$. The closest minimum value corresponds to the next zero value of the first derivative $\dot{x}(\Delta t) = A_1\omega_1 \sin \omega_1 \Delta t + A_2\omega_2 \sin (\omega_2\Delta t) = 0$. The last nonlinear equation can be solved analytically by $\Delta t = -\omega_1^{-1} \arcsin[\frac{A_2\omega_2}{A_1\omega_1} \sin(\omega_2\Delta t)]$, if $A_1\omega_1 \geq A_2\omega_2$, $\Delta t = -\omega_2^{-1} \arcsin [\frac{A_1\omega_1}{A_2\omega_2} \sin(\omega_1\Delta t)]$, if $A_1\omega_1 < A_2\omega_2$. The obtained solution of the time moment for the closest minimum value $\Delta t = S_1(A_2/A_1, \omega_2/\omega_1)$ depends only on the relations between the amplitudes and frequencies of the harmonics.

This solution makes it possible to generate the closest minimum values from left and right and, since the cosine is the even function $(x_{\min}(\Delta t) = x_{\min}(-\Delta t))$, the virtual synchronous minimum value will be: $x_{\min}(0) = A_1 \cos \omega_1 \Delta t + A_2 \cos (\omega_2\Delta t)$. The arithmetic average of the initial maximum and the obtained synchronous minimum can be written in the form $F_1 = 0.5 [x_{\max}(0) + x_{\min}(0)] = 0.5 [A_1 + A_2 + A_1 \cos \omega_1 \Delta t + A_2 \cos (\omega_2\Delta t)]$.

It is convenient to divide the obtained solution into two parts and analyze them separately, initially showing only the first harmonic modification (Figure 6.6a)

$$F_{1,1} = 0.5A_1 [1 + \cos \omega_1 \Delta t] \tag{6.4}$$

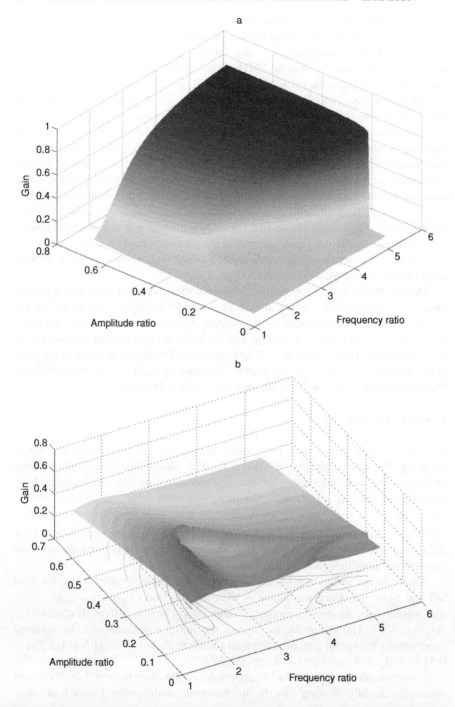

Figure 6.6   The theoretical mean value between the local maxima and minima at the highest maximum position: the envelope of the first harmonic (a), the envelope of the second harmonic (b)

and then the second part with the second harmonic modification (Figure 6.6b)

$$F_{1,2} = 0.5A_2 \left[1 + \cos\left(\omega_2 \Delta t\right) \right]. \tag{6.5}$$

Each of these parts describes an arithmetic mean value between the top and bottom extrema in the highest current position of the local maximum when $t_i = 0$, $\varphi = 0$.

The case of $t_i = \pi$, $\varphi = \pi$

This is a case in which the initial maximum value is equal to $x_{max}(\pi) = -A_1 + A_2$ and the initial value of the first derivative is equal to $\dot{x}(\pi) = 0$. The closest minimum value corresponds to the next zero value of the first derivative $\dot{x}(\Delta t) = -A_1\omega_1 \sin \omega_1 \Delta t + A_2\omega_2 \sin(\omega_2 \Delta t) = 0$. The solution of this last nonlinear equation determines the moment in time for the closest minimum value $\Delta t = S_2(A_2/A_1, \omega_2/\omega_1)$, which depends only on the relations between the amplitudes and frequencies of the harmonics.

The solution allows us to generate the closest minimum values from both left and right, and, since the cosine is an even function $(x_{min}(\Delta t) = x_{min}(-\Delta t))$, the virtual synchronous minimum value will be: $x_{min}(\pi) = -A_1 \cos \omega_1 \Delta t + A_2 \cos(\omega_2 \Delta t)$. The arithmetic average of the initial maximum and the obtained synchronous minimum take the form $F_2 = 0.5[x_{max}(\pi) + x_{min}(\pi)] = 0.5[-A_1 + A_2 - A_1 \cos \omega_1 \Delta t + A_2 \cos(\omega_2 \Delta t)]$.

Again, it is convenient to divide the obtained solution and analyze the two parts separately, with the first part showing only a modification of the amplitude of the first harmonic (Figure 6.7a)

$$F_{2,1} = \left| -0.5A_1 \left(1 + \cos \omega_1 \Delta t \right) \right| \tag{6.6}$$

and the second part showing only a modification of the amplitude of the second harmonic (Figure 6.7b)

$$F_{2,2} = 0.5A_2 \left[1 + \cos\left(\omega_2 \Delta t\right) \right]. \tag{6.7}$$

Each part describes the arithmetic mean value between the top and bottom extrema in the lowest current position of the local maximum when $t_i = \pi, \varphi = \pi$.

6.2.5   EMD as a nonstationary and nonlinear filter

The final step of the EMD algorithm subtracts the obtained arithmetic mean function of time from the initial signal. Such subtraction generates the simplest components known as the IMF. Therefore, the subtraction procedure specifies a digital filtering operation where the input is the initial signal composition; the output is the IMF; and the filter characteristics (6.4)–(6.6) are represented in Figures 6.6 and 6.7.

In order to understand how the IMF is extracted from the initial signal, let us analyze the obtained filter characteristics. This analytical three-dimensional filter is defined as a 3D function with two arguments: the relative harmonics amplitude ratio

a

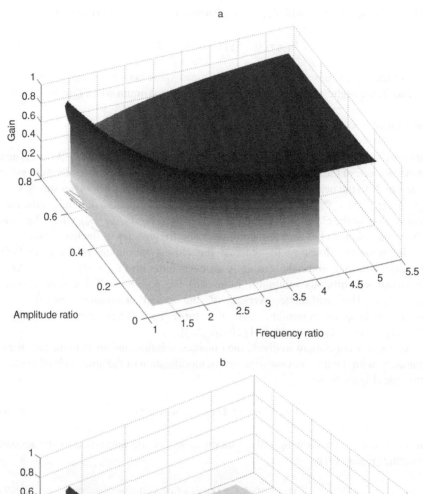

b

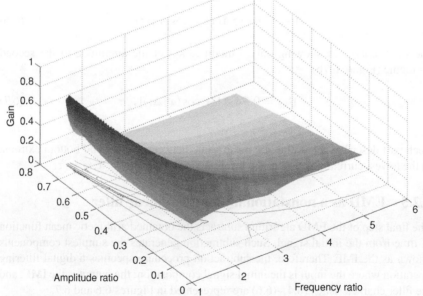

Figure 6.7   The theoretical mean value between the local maxima and minima at the lowest maximum position: the envelope of the first harmonic (a), the envelope of the second harmonic (b)

A_2 ($A_1 = 1$) and the relative harmonics frequency ratio ω_2 ($\omega_1 = 1$). The vertical value of the surface presents a portion of the magnitude passing through the filter. Vertical values that are close to 1 indicate that the output signal passes through the filter, while those that are close to 0 indicate a rejection of the output signal. The obtained analytical solutions demonstrate that the EMD is nonstationary and nonlinear at the final step of estimating the IMF. As the EMD is nonstationary, this means that the filter characteristics are varied, with the extremum roaming from the top to the bottom position. The nonlinearity is embedded in the filter magnitude dependency on the harmonics amplitude and frequency ratio.

The main common property of the obtained analytical filters for both the top and bottom extrema positions is their high magnitude values for the first harmonic when $\omega_2 \to \infty$ and $A_2 \to 1$. This is a tendency of the highpass frequency to get through the unmodified first harmonic with a larger frequency and amplitude ratio.

Another common property of the analytical filters is their highpass magnitude values for the first harmonic and small (almost rejection) magnitude values for the second harmonic. As a result, the low-frequency first harmonic $A_1 \cos \omega_1 t$ can fully pass through the filter, while the high-frequency second harmonic will be stopped. Then, after being subtracted from the initial composition, the low-frequency harmonic will disappeared, and the final IMF will consist only of the second high-frequency harmonic $A_2 \cos \omega_2 t$. One more important common property of the analytical filters is the existence of a separating boundary surface B ($A_2, \omega_2, A_1 = 1, \omega_1 = 1$) dividing the space of parameters into two ranges, thus allowing (or not allowing) the low-frequency harmonic to pass.

6.2.6   Frequency resolution of the EMD

The displayed separation boundaries are directly related to the frequency resolution characteristics of the EMD. For more precise analysis, let us plot the 2D projection of the same nonlinear filters with the axes $A_2, \omega_2, (A_1 = 1, \omega_1 = 1)$ as pseudo-color graphs (Figure 6.8). The intensities of all function variations and nuances are usually much better defined in such a 2D graph. Depending on the amplitude and frequency ratio, the limiting boundary determines the region to the right where the EMD is able to separate harmonics, and the region to the left where the EMD cannot separate two tones. The top extrema position filter (Figure 6.8a) has a gentle slope and a cutoff boundary that runs in the direction of higher frequency from the right side. A good power approximation of the boundary (Figure 6.8a, dashed line) shows that the filter blocks the first harmonic when $(A_2/A_1)_{\text{boundary(top)}} \leq 1.44(\omega_2/\omega_1)^{-1.4}$, and passes the first harmonic without modification for higher relations $A_2/A_1 > (A_2/A_1)_{\text{boundary(top)}}$. In the same figure (Figure 6.8a, thick line) we plotted the curve corresponding to the inversely related amplitude and frequency ratio $A_2/A_1 = \omega_1/\omega_2$ for the case of a negative IF (4.10). This inversely related amplitude and frequency ratio is fairly close to the theoretical boundary curve.

The bottom extrema position filter (Figure 6.8b) has a more aggressive slope. Its cutoff boundary is shifted to the left, and the filter blocks the first harmonic exactly when $(A_2/A_1)_{\text{boundary(bottom)}} \leq (\omega_2/\omega_1)^{-2}$. For the harmonics ratios located on the left of the aggressive bottom filter boundary, the filter does not allow the EMD to extract

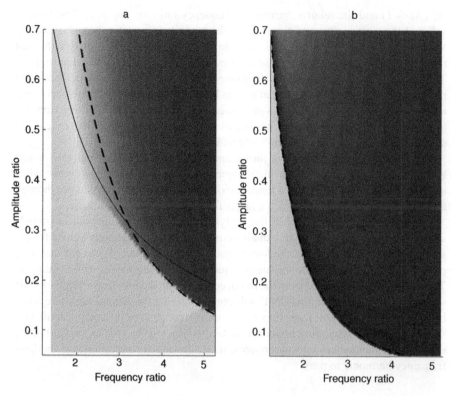

Figure 6.8   The theoretical boundary of the first harmonic filtering: the highest max-imum position (a), the approximation $A_2/A_1 \leq 1.44\,(\omega_2/\omega_1)^{-1.4}$ (- -), the approximation $A_2/A_1 = \omega_1/\omega_2$ (–); the lowest maximum position (b), the approximation $A_2/A_1 \leq (\omega_2/\omega_1)^{-2}$ (- -)

the harmonics at all, no matter how many sifting iterations are involved. It is evident that the harmonics ratios located on the right of the gentle top filter boundary allow the EMD to extract the harmonics completely and at once during the first iteration.

In the case where the harmonics frequency intervenes between these boundaries, the first harmonic will pass the filter partially at every sifting iteration, with the attenuation coefficient depending on the filter slope. To approach the full value of the envelope, the first harmonic should be passed through the filter several times. In other words, harmonics whose frequency ratio is located between these theoretical boundaries might be separated with several sifting iterations.

Logically, the presented analytical nonstationary and nonlinear filters describe the EMD filtering capacity only in extreme positions. In reality, the current positions of the extrema change constantly and the filters are continuously being transformed from one to another, producing a kind of mid-position filtering. The concrete spline-fitting algorithm used in the real EMD program can also have some impact on the filter's slope, but both obtained extreme analytical boundaries will remain unchanged.

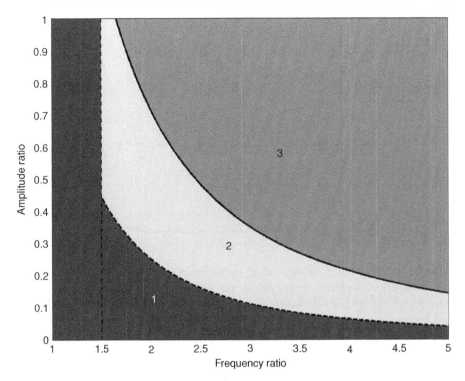

Figure 6.9   The EMD ranges of two harmonics separation: (1) the impossible decomposition for very close frequency harmonics and small amplitude ratio; (2) the decomposition requires several sifting iterations for close frequency harmonics; (3) the single iteration separation for distant frequency harmonics and large amplitude ratio

That is, the more the frequencies are spaced apart, the less is the amplitude ratio of two harmonics suitable for EMD separation. For example, a second harmonic with a tripled or lower frequency $\omega_2 \leq 3\omega_1$ and a small amplitude less than $A_2 < 0.3A_1$ can be extracted with some iterations. Nevertheless, any amplitude that is less than $A_2 \leq 0.11A_1$ absolutely cannot be separated by the EMD. For example, if frequencies lie within an octave of each other $f_2 \leq 2f_1$ and their amplitudes differ by less than $A_2 \leq 0.25A_1$, the EMD method is unable to separate two such components. This means that the EMD does not perform well for smaller amplitudes of the second harmonic and cannot distinguish frequencies that are close to each other

The logical result of the provided theoretical analysis is that the frequency ratio of the harmonics can be assembled into three different groups (Figure 6.9): (1) very close frequency harmonics $A_2/A_1 \leq (\omega_1/\omega_2)^2$ unsuitable for EMD decomposition (Figure 6.10), (2) close frequency harmonics $(\omega_1/\omega_2)^2 \leq A_2/A_1 < 1.44\,(\omega_1/\omega_2)^{1.4}$ requiring several sifting iterations (Figure 6.11), and (3) distant frequency harmonics $A_2/A_1 > 1.44\,(\omega_1/\omega_2)^{1.4}$ that are well separated by a single iteration (Figure 6.12).

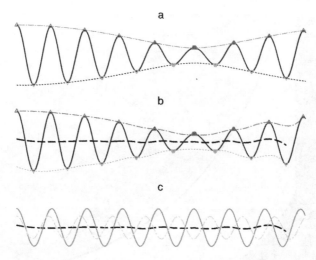

Figure 6.10   The EMD of two very close harmonics ($A_1 = 1$, $\omega_1 = 1$, $A_2 =$ 0.6, $\omega_2 = 1.1$): the initial signal (a, –), the upper (···) and lower (···) envelope, the top ($\triangle$) and bottom maxima points (∇); the initial signal (b, –), the top (b, ···) and bottom (b, ···) extrema curves, the mean value between them (b, - -); the first harmonic (c, –), the second harmonic (c,- -), the mean value between the top and bottom extrema (c, - -)

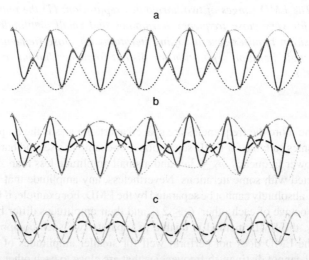

Figure 6.11   The EMD of two close harmonics ($A_1 = 1$, $\omega_1 = 1$, $A_2 = 0.9$, $\omega_2 =$ 1.8): the initial signal (a, –), the upper (···) and lower (···) envelope, the top ($\triangle$) and bottom maxima points (∇); the initial signal (b, –), the top (b, ···) and bottom (b, ···) extrema curves, the mean value between them (b, - -); the first harmonic (c, –), the second harmonic (c,- -), the mean value between the top and bottom extrema (c, - -)

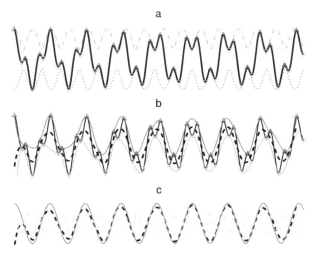

Figure 6.12   The EMD of two distant harmonics ($A_1 = 1$, $\omega_1 = 1$, $A_2 = 0.4$, $\omega_2 = 2.9$): the initial signal (a, –), the upper (---) and lower (···) envelope, the top (Δ) and bottom maxima points (∇); the top (b, –) and bottom (b, –) maxima curves, the mean value between them (b, - -); the decomposed first harmonic (c, –), the second harmonic (c,- -), the mean value between the top and bottom maxima curves (c, - -)

The first group corresponds to the always positive IF of the composed signal. The case is a well-known type – beating oscillation. An example of beating oscillation induced by two close harmonics is shown in Figure 6.10 as well as a varying envelope and an opposite sign envelope. Both envelopes alternate slowly with the average frequency $\omega_2 - \omega_1$. Notice, that the initial signal alternates much faster with the average frequency ω_1. Every discrete local extreme point of the initial signal "samples" the corresponding essentially continuous envelope. Connecting all consecutive discrete local maxima yields a set of another function – the upper extreme line; connecting all consecutive local minima yields the lower extreme line (Figure 6.10a). These slow-varying extreme lines akmoxt coincide with their corresponding envelopes. It is clear that computing a mean value between such close upper and lower extreme lines will produce zero, because the envelope and the opposite sign envelope are always out of phase. This indicates that the EMD method cannot decompose close frequencies. The same negative result is true regarding any AM signal (Section 4.5) that cannot be decomposed by the EMD method.

6.2.7   Frequency limit of distinguishing closest harmonics

According to (6.1), the continuous function of the cosine projection $x_{extr}(t)$ oscillates with the frequency $\omega_2 - \omega_1$ posed by the envelope oscillation. This function is sampled with a frequency equal to the IF $\omega(t)$ to form extrema points. If the frequency of the oscillation is larger than the Nyquist frequency $\omega_2 - \omega_1 > 0.5\omega(t)$, the sampled extrema curves undergo a nonlinear filtering, as described in the previous sections. If the frequency of the oscillation is less than the Nyquist frequency

$$\omega_2 - \omega_1 \leq 0.5\omega(t), \tag{6.8}$$

then no aliasing (folding) occurs, and the resultant sampled extrema curve will oscillate with the same frequency $\omega_2 - \omega_1$ as the initial envelope. Such retention of frequency means that the top extrema curve will repeat the upper envelope and, correspondingly, the bottom extrema curve will repeat the lower envelope. As a result, the average of these extrema curves will always produce zero, and the EMD will not be able to decompose harmonics.

The simple formula (6.8) provides a strong limit on the possibility of the EMD method being able to operate. According to (5.3), the average value of the IF of the composition is equal to the frequency of the largest harmonics. In our notations the average frequency of two harmonics is always $\overline{\omega}(t) = \omega_1$. Substituting this value in (6.8) yields $\omega_2 - \omega_1 \leq 0.5\omega_1$ and, finally,

$$\omega_2 \leq \frac{3}{2}\omega_1 \tag{6.9}$$

Equation (6.9) yields the smallest value of the second harmonic frequency that the EMD is able to distinguish. If the value of ω_2 is any lower, the EMD is unable to separate the components. This smallest value of ω_2 is an absolute strong limit that does not depend on the amplitude relations between harmonics. This theoretical limit value completely coincides with the experimental critical frequency ratio $\omega_1/\omega_2 \approx 0.67$ found in the work of Rilling and Flandrin (2008). Above this value, it is impossible to separate the two components, no matter what the amplitude ratio. This is the case when the local extrema do not differ from the corresponding envelope curves.

Using the first derivative of the signal in the signal analytic form, we devised an expression for the local extremum points, including their vertical locations and distribution in time. As shown above, the obtained vertical position of the local extrema can deviate from the envelope, thus explaining, for instance, why and when the maximum points can become negative.

The displayed extrema deviation from the envelope forms the basis for a theoretical explanation of the EMD sifting procedure. A vertical distance between the envelope and the local extrema depends on the relation between the first derivative of the envelope from one side and the multiplication of the envelope by the IF from the other. To build the simplest synchronous extrema, we suggest connecting the opposite closest neighboring left and right extrema, thus yielding a theoretical median function between the top and bottom extrema of two harmonics. This theoretical median function represents a sort of signal nonlinear filter whose input is an initial two-tone composition and whose output can be a harmonic with the lowest frequency. Depending on the amplitude and frequency ratios of the harmonics, the filter passes through some portion of the magnitude of the lowest frequency. The filter is nonstationary because its characteristics vary, while the extrema roam from the highest to the lowest position. At these extreme positions, the filter characteristics differ: at the highest position, the filter has a gentle slope, but at the lowest position it has a more aggressive slope.

The obtained boundaries between the filter pass and the stop characteristics determine the resolution of the EMD theoretical frequency. When the smaller harmonic amplitude is less than the boundary of the aggressive slope, the EMD does not separate the harmonics. When the smaller harmonic amplitude is larger than the boundary of the gentle slope, the EMD separates the harmonics according to its first single sifting iteration. Middle amplitudes between the boundaries require several iterations, depending on the filter attenuation pass characteristics. This way the initial composition – after extracting the median function – will contain a high-frequency harmonics, such as the intrinsic mode function. This explains how the EMD uses sifting to decompose the first high-frequency components. For two-tone models, the critical frequency limit of distinguishing the closest harmonics was found theoretically. The harmonics with a frequency below critical cannot be extracted from the decomposition by the EMD, no matter how large its amplitude.

Like any other signal-processing procedure, the EMD operates with an input signal only. The EMD decomposes the signal exclusively by means of its inherent organic transformation function. For a composition of harmonics it extracts the highest frequency of the composition first. Like any other signal analysis instrument, it merely reflects and represents real physical and natural processes and phenomena. It is incorrect to try to understand or explain the EMD tool through nonlinear structural dynamics or through any other physics-based foundation.

6.3   Global Hilbert Vibration Decomposition

The same problem of decomposition of nonstationary and multicomponent vibration can be solved by a different technique, called the Hilbert Vibration Decomposition (HVD) method (Feldman, 2006). The HVD method is based theoretically on the HT presentation of the IF and does not involve spline fitting and empirical algorithms. A principle of the proposed HVD method is to decompose the initial vibration $x(t)$ into a sum of components with slow-varying instantaneous amplitudes and frequencies. Such an identification of every inherent synchronous component belonging to different time scales can be made on the basis of the global time domain analysis of the IF of the initial signal. It is natural that each of the inherent synchronous components must have physical and mathematical significance. For better understanding of the meaning of the global HVD method for the separation of vibration components by the HT, we examine some mathematical issues.

6.4   Instantaneous frequency of the largest energy component

A simple case of a combination of two harmonics $x(t) = A_1 \cos \omega_1 t + A_2 \cos \omega_2 t$ (Section 4.8) enables us to discover and prove an important feature of the IF. Equation (4.9) shows that the IF consists of two different parts – that is, a slow-varying frequency of the first component ω_1 and a rapidly varying asymmetrical oscillating part. We already proved that integration of the oscillating part of the IF results in zero

$$\int\limits_0^T \frac{(\omega_2 - \omega_1)\left[A_2^2 + A_1 A_2 \cos\left(\int (\omega_2 - \omega_1)dt\right)\right]}{A^2(t)} dt = 0; \; A_1 > A_2. \qquad (6.10)$$

This means that the average value or the first moment of the IF is just equal to the frequency of the largest harmonic $\langle \omega(t) \rangle = \omega_1(t) + \int_0^T \omega(t) = \omega_1(t) + 0$. If there are three or more quasiharmonics in the composition, the IF will take a more complicated form, but again, the averaging or the lowpass filtering will extract only the IF of the largest energy component.

This important property of the IF offers the simplest and most direct way of estimating the frequency of an *a priori* unknown largest synchronous signal component. Instead of integration, we will use averaging by a convolution of the IF with a lowpass filter which will cut down the fast asymmetrical oscillations and leave only a slow-varying frequency of the main signal component $\omega_1(t)$. A cutoff frequency of the lowpass filter that divides the passband and the stopband will control the frequency resolution of the HVD method. In other words, the signal components can be separated only when the difference between their frequencies is more than the value of the cutoff frequency. A simple lowpass filtering of the IF leads to the estimation of the frequency of the largest energy signal component. The next step is to estimate the envelope of the largest energy signal component.

6.5   Envelope of the largest energy component

To estimate the envelope for the recognized frequency we choose a well-known technique (Fink, 1975) known variously as: synchronous detection, in-phase/quadrature demodulation, coherent demodulation, autocorrelation, signal mixing and frequency shifting, lock-in amplifier detection, and phase sensitive detection. In essence, the technique extracts the amplitude details of a varying synchronous vibration component with a known frequency by multiplying the initial vibration composition by two reference signals exactly 90° out of phase with one another (see Section 3.2). For the output we will get two projections: the in-phase and the HT (quadrature) phase output. The vibration amplitude can be obtained by lowpass filtering of these projections and taking the square root of the sum of their squares. It is desirable to choose the cutoff frequency value as small as possible, but not less than the frequency value of the lowest vibration component.

In this case, a single synchronous vibration component $x_{l=r}(t) = A_{l=r}(t) \cos\left(\int \omega_{l=r}(t)dt\right)$ with exactly the same frequency as the reference vibration $\cos\left(\int \omega_r(t)dt\right)$ is mixed with the other l-vibration components. The synchronous detection technique is capable of measuring even small varying vibrations that are obscured by large numbers of other components.

6.6   Subtraction of the synchronous largest component

As a result of two suggested procedures, we estimated the IF as well as the envelope of the first synchronous largest vibration component $x_1(t) = A_1(t) \cos \left(\int \omega_1(t) dt \right)$. The first largest demodulated signal component is estimated during the first step of the iteration of the HVD method. Using the idea of signal sifting (Huang *et al.*, 1998), we can subtract the synchronous largest component from the initial composition $x_{l-1}(t) = x(t) - x_1(t)$, thus obtaining a new signal composition $x_{l-1}(t)$ that could be decomposed again during the next iteration. At every iteration step the residual signal will contain the information on lower energy components. This way we divide the initial composition into several slowly varying oscillating components. A criterion for stopping the sifting process can be the required number of components or the limit value of the standard deviation difference computed from two consecutive iteration results.

Contrary to the well-known EMD method, which starts with the first intrinsic mode with the highest frequency component, the proposed HVD method starts decomposition with the synchronous largest energy component while the residual signal contains information of lower energy components. As a result, the first synchronous component separated from the initial signal according to the HVD method contains the highest varying amplitude. The residual signal contains information on other lower amplitude components. Naturally, each of the components must have physical and mathematical significance.

During the first step of the iteration for the proposed HVD method, we found the synchronous largest vibration component. According to the basic principles of demodulation, we use a well-known synchronous detection technique, multiplying the carrier signal component by the modulating signal. In this case, a single oscillating component $x_{l=r}(t) = A_{l=r}(t) \cos \left(\int \omega_{l=r}(t) dt \right)$ of exactly the same varying frequency as the reference signal $\cos \left(\int \omega_r(t) dt \right)$ is mixed with other l-components. In essence, the technique extracts the amplitude details about a synchronous oscillation component with a known frequency by multiplying the initial composition by two reference signals exactly 90° out of phase with one another. We defined each decomposed term as a vibration in unison of the composition (i.e., a *synchronous component*). This way every particular component of a nonstationary signal is characterized by the synchronous evolution of the total composition.

Oscillating components that are not of the exact same frequency as the reference ($\omega_l \neq \omega_r$) will not yield this slow-varying function. No matter what the instantaneous phase, the resultant envelope $A_{l=r}(t)$ always represents the detected synchronous component envelope. The synchronous detection technique is capable of measuring even small varying signals that are obscured by large numbers of other components.

Using the idea of signal sifting (Huang *et al.*, 1998), we subtract the largest component from the initial composition $x_{l-1}(t) = x(t) - x_1(t)$, thus obtaining the new vibration composition $x_{l-1}(t)$ that should be decomposed again during the next iteration $x_2(t) = A_2(t) \cos \left(\int \omega_2(t) dt \right)$. As a result, we split the initial composition into several slow-varying vibration components. The number of iterations necessary

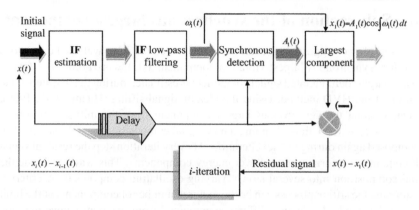

Figure 6.13   Block diagram of the HVD method

to provide a good approximation to a vibration composition depends on how rapidly the initial vibration changes.

6.7   Hilbert Vibration Decomposition scheme

The idea of the HVD method is to decompose an initial wideband oscillation $x(t)$ into a sum of components with slow-varying instantaneous amplitudes and frequencies, so that $x(t) = \sum_l A_l(t) \cos \left(\int \omega_l(t) dt \right)$, where $A_l(t)$ is the instantaneous amplitude and $\omega_l(t)$ is the IF of the l-component. The decomposition is based on the assumptions that (1) the underlying signal is formed by a superposition involving at least one quasiharmonic function with several full period lengths; and (2) the envelope and frequency of each oscillating component differ.

The proposed decomposition is an iterative method, and every iteration step includes the following three procedures (Figure 6.13): (a) an estimation of the IF of the largest component; (b) a detection of the corresponding envelope of the largest component; and (c) a subtraction of the largest component from the composition. The key factor of a precise decomposition is to use appropriate methods to extract the IF and envelope of the initial vibration composition.

On each iteration step, the corresponding slow-varying synchronous vibration component is extracted by the lowpass filtering of the IF. A lowpass filter will eliminate all high-frequency components outside of cutoff frequency and allow all the low-frequency components to pass without modification.

A spectral content of a vibration signal can be represented by modulation of its instantaneous frequency. Passing the signal through a narrowband frequency filter gets rid of the distant side frequencies and makes the instantaneous frequency less and more slowly modulated. So the narrowband signal filtering is equivalent to the lowpass filtering of the instantaneous frequency. In this case the width of the narrowband filter corresponds to the cutoff frequency of the lowpass filter.

6.7.1   Frequency resolution of the HVD

A cutoff frequency that divides the passband and the stopband of the lowpass filter will control the frequency resolution of the HVD method. It is desirable to choose a cutoff frequency value that is as small as possible considering the filter shape factor accuracy and stability. On each iteration step, after subtracting the previous frequency, the current frequency becomes the next in term of the lowest frequency, and again its value should be larger than the cutoff frequency value. This means that the difference between close frequencies should be larger than the value of the cutoff frequency. In other words, the vibration components can be separated only when the difference between their frequencies is more than the value of the filter cutoff frequency.

For a more precise frequency resolution the cutoff frequency of the lowpass filter should be as small as possible. Typically, the smallest cutoff frequency value of a stable and precise lowpass filter is $f_{min} \geq 0.02Fs$, where Fs is the sampling frequency. For a harmonic with a frequency f_i that is sampled, say, with twenty points per period ($f_i = 0.05F_s$), the filter can produce the next higher distinguishing frequency equal to $f_{i+1} = (0.05 + 0.02)Fs = 0.07Fs$. As a result, the frequency components will differ by $0.07/0.05 = 1.4$, and several frequency components (0.05, 0.07, 0.09) lying in the same octave can be separated. The frequency resolution can be improved even further by decreasing the sampling frequency.

The frequency resolution for the HVD method does not depend on a dissimilar harmonics amplitude ratio. Therefore the graphic representation of the frequency resolution shows the frequency value relative to the sampling frequency (Figure 6.14).

6.7.2   Suggested types of signals for decomposition

The proposed method is dedicated primarily to quasi and almost periodic oscillating-like signal decomposition. Such a vibration type could be, for example, an amplitude and frequency modulated vibration, a nonstationary vibration similar to that at rotor startup or shutdown; or a nonlinear dynamic system vibration. It basically applies the amplitude and frequency modulated harmonics instead of the constant ones used by the Fourier transform. The HVD method cannot separate other types of motion, such as random, impulse, short or nonoscillating (aperiodic) signals. Nor is it intended for the case of very closely spaced frequencies. The HVD method is illustrated further using the data from numerical simulation results published on the Internet as the MATLAB pre-parsed pseudo-code files (P-file) (Feldman, 2008a).

The HVD method, as opposed to other known decomposition methods, is extremely simple and fast in calculation. An additional – and very useful – feature is its ability to detect synchronous vibration components with desirable or specified frequencies – for example, those with only odd high harmonics of the main vibration component.

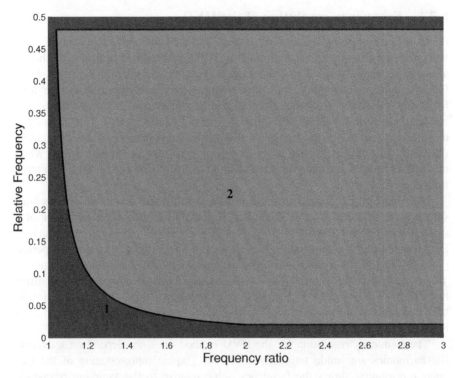

Figure 6.14   The HVD ranges of two harmonics separation: (1) impossible decomposition for very close frequency harmonics; (2) good separation for distant frequency harmonics

6.8   Examples of Hilbert Vibration Decomposition

To get an idea of the HVD method, it is instructive to find elementary components for some examples of nonstationary vibration signals with varying amplitude and frequency.

6.8.1   Nonstationary single-sine amplitude modulated signals

An AM signal is generated by modulating an amplitude of a carrier signal; it is composed of two terms: an oscillating carrier wave plus a wave that is a product of two sinusoidal terms. Actually, the AM signal is composed of three sinusoidal waveforms: a carrier wave, a lower sideband signal, and an upper sideband signal. In the general case, the AM signal can be modulated with a nonstationary modulating function.

For illustration purposes, let the carrier signal be a harmonic function $\cos t$, and let the modulation also be a single, but nonstationary, tone with a decreasing modulation index and an increasing modulated frequency: $x(t) = [1 + \frac{t_{max}-t}{t_{max}} \cos(\frac{t^2}{4t_{max}})] \times \cos t; \quad t = [0 \ldots 1024]$. The signal waveform is shown in Figure 6.15a, and the

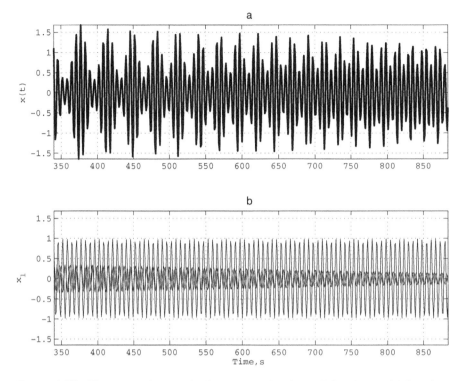

Figure 6.15   The nonstationary single-tone amplitude modulated signal (a) and its decomposed superimposed components (b) (Feldman, ©2009 by John Wiley & Sons, Ltd.)

synchronous components decomposed according to the HVD method are shown in Figure 6.15b. The same three decomposed components are shown separately in Figure 6.16, where the first carrier component has a unit amplitude and a constant frequency, while the nonstationary sideband components vary in time. The increasing modulated frequencies that are symmetrically located around the carrier frequency and the decreasing modulation amplitudes are depicted separately in Figures 6.17 and 6.18, respectively, in more detail. Thus, the proposed method illustrates some known important properties of a single-tone AM signal.

6.8.2   Nonstationary overmodulated signals

In practice, existing demodulation techniques always recommend that the AM signal $x(t) = [A_0(t) + A_m(t) \cos \omega_m t] \cos \int \omega_0(t)dt$ should have a modulation index smaller than unity $m = A_m/A_0 < 1$ (see Section 4.5), because it is difficult to demodulate such an overmodulated nonstationary signal with a complex envelope, not to mention the duality relations for the alternate envelope $1 + m \cos \omega_m t$ which can have both positive and negative values (Cohen and Loughlin, 2003; Wetula, 2008).

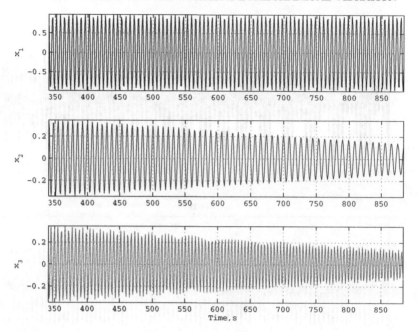

Figure 6.16   First three components of a nonstationary single-tone amplitude mod-ulated signal

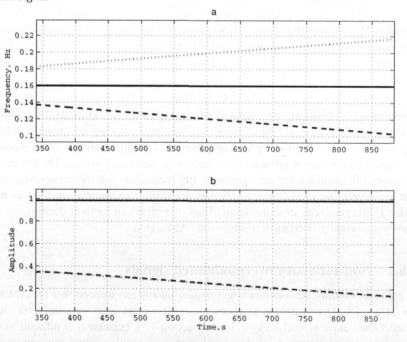

Figure 6.17   The IF (a) and envelope (b) of a single-tone amplitude modulated signal: the carrier signal component (–), the low (- - -) and high (⋯) modulation component

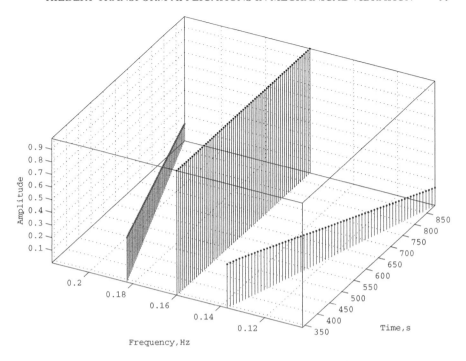

Figure 6.18   The Hilbert spectrum of the single-tone amplitude modulated signal

Therefore, for the existing techniques, overmodulation is almost always considered to be a fault condition.

The distinctive property of the HVD method is its ability to decompose the overmodulated signal even when a carrier component has a very low value. By applying the HVD method we will decompose three synchronous components of the AM signal $x(t) = [A_0(t) + A_m(t) \cos \omega_m t] \cos \int \omega_0(t)dt = x_1(t) + x_2(t) + x_3(t)$, where one component $x_1(t) = A_0(t) \cos \int \omega_0(t)dt$ is the low amplitude carrier wave, and two others $x_2(t) = \frac{A_m(t)}{2} \cos \int [\omega_m(t) - \omega_0(t)] \, dt$ and $x_3(t) = \frac{A_m(t)}{2} \cos \int [\omega_m(t) + \omega_0(t)]dt$ are the large and equal amplitude sidebands. The half-sum of two extracted instantaneous frequencies produces the frequency of the alternate envelope $05 [\omega_m(t) - \omega_0(t) + \omega_m(t) + \omega_0(t)] = \omega_m(t)$. The double envelope of the extracted equal sidebands produces the magnitude of the alternate envelope. Thus, the alternate envelope can be estimated according to the following expression:

$$A_\pm(t) = A_0(t) + A_m \cos \int \omega_m(t)dt. \qquad (6.11)$$

As an illustration, let us take a nonstationary overmodulated amplitude signal $x(t)$ whose modulation function $A_0(t) + A_m(t) \cos \omega_m t$ has negative values: $x(t) = \{1 + [6 + 2 \sin(0.02t)] \cos(0.1 + 0.1 \frac{2t_{max} - t}{t_{max}})t\} \times \cos t; \quad t = [0 \ldots 1000]$. The signal is shown in Figure 6.19a, along with the positive envelope function. The

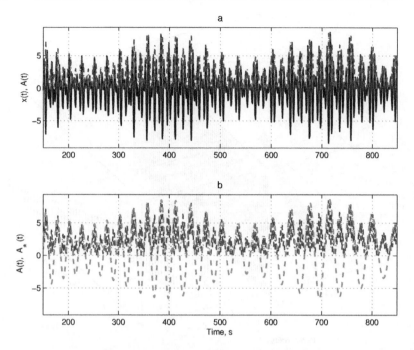

Figure 6.19   The overmodulated AM signal (a, —), its envelope (— -□-); the alternate envelope (b, - -) (Feldman, ©2011 by Elsevier)

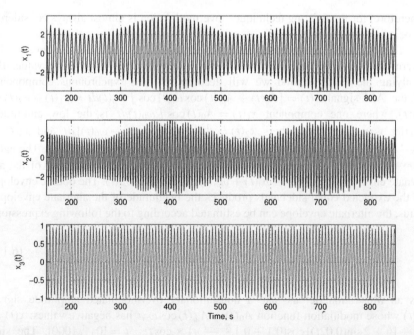

Figure 6.20   The first three components of a nonstationary overmodulated AM signal (Feldman, ©2011 by Elsevier)

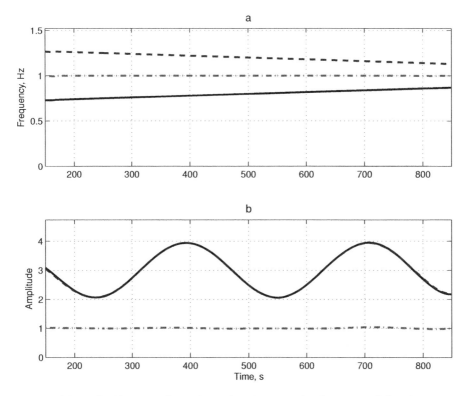

*Figure 6.21   The IF (a) and envelope (b) of an amplitude overmodulated signal;
the carrier signal component (···), the low (–) and high frequency (- -) modulation
component (Feldman, ©2011 by Elsevier)*

decomposed components are shown separately in Figure 6.20. The envelope and
the IF of each component – presented in Figure 6.21 and in Figure 6.22 as the Hilbert
spectrum – demonstrate the nonstationary variation of the modulation function, while
the varying envelopes of both sidebands are the same. These figures also show that
the varying sideband amplitude is higher than the carrying amplitude. The obtained
alternate envelope is shown in Figure 6.19b with the positive envelope. It is clear that
the alternate envelope of the overmodulated signal is much simpler than the positive
envelope. The choice between the two possible forms of envelope representation
depends on the operability and simplicity of further operations with the signal.

6.8.3   Nonstationary waveform presentation

A good approximation of the sum of decomposed synchronous components to the
initial signal depends on how rapidly the initial waveform is changing and on the total
number of obtained components that is equal to the number of HVD iterations. As
an example, Figure 6.23a (dashed line) depicts the signal of the nonstationary square

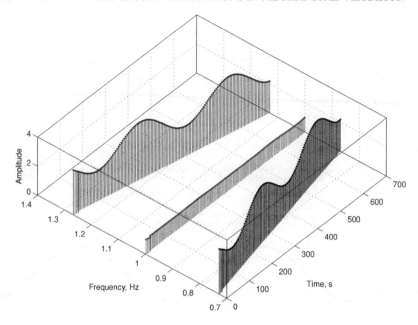

Figure 6.22   The Hilbert spectrum of an amplitude overmodulated signal (Feldman,
© *2011 by Elsevier)*

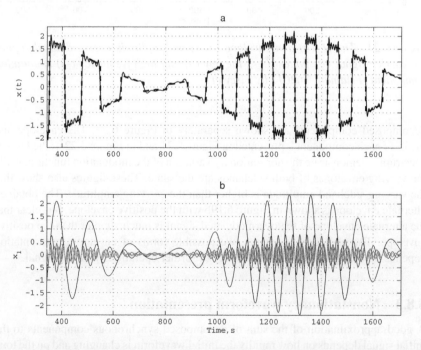

Figure 6.23   The nonstationary square wave: the initial signal (a, - -), the sum of the
first five components (a, −), and the decomposed superimposed components (b)

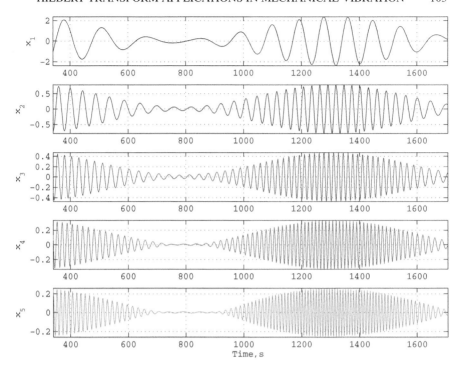

Figure 6.24   The first five components of a nonstationary square wave

wave function with varying amplitude and period $x(k) = (1 + 0.9\sin(0.006k)) \times$
$\text{sgn}[\sin(0.05 + 0.04\sin(0.005k)k)]$, $k = [0..2048]$.

The first five decomposed synchronous component terms are plotted as a sum together with the initial square wave in Figure 6.23a (solid line). As can be seen, just the first five components describe the nonstationary square wave in the time domain with a high degree of accuracy. Further details about each synchronous oscillation component can be found in Figure 6.23b, and also in Figure 6.24 where they are shown separately. Figures 6.25 and 6.26 present the IF and the envelope of each component of the nonstationary square as time-varying functions.

This example illustrates the fact that the proposed HVD method makes it possible to construct a perfect match between a complicated nonstationary signal and a composition of the small number of time-varying elementary oscillating components. The results of the HVD describe a nonstationary square wave in the time domain with a high degree of accuracy.

6.8.4   Forced and free vibration separation

To illustrate the potentials of the proposed decomposition method, consider the case of a nonstationary vibration (involving a steady-state and a transient component) generated by a chirp force. The following numerical example includes the linear

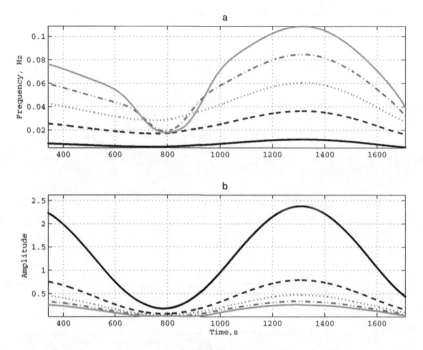

Figure 6.25   The IF (a) and envelope (b) of each component of the nonstationary square wave: the first signal component (–)

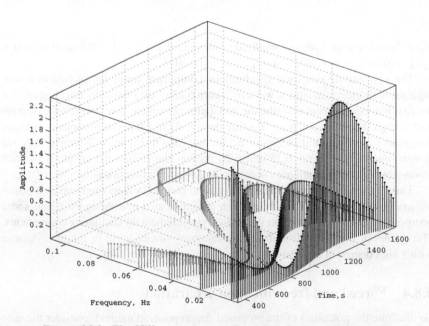

Figure 6.26   The Hilbert spectrum of a nonstationary square wave

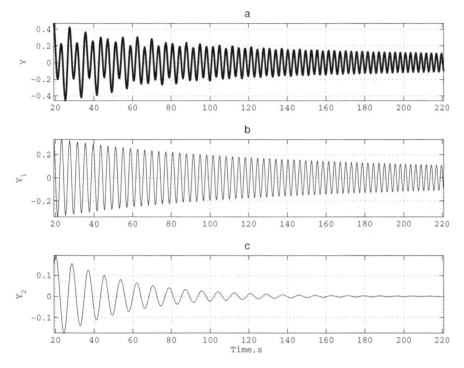

Figure 6.27   The nonstationary vibration solution (a) and the separated vibration components: the steady state (b), the transient (c)

dynamic system excited by a sine wave whose frequency increases over time at a linear rate $\ddot{x} + 0.07\dot{x} + x = \cos(6.5 \times 10^{-5}t^2)$.

Figure 6.27a shows the initially forced vibration resulting from the application of an external periodic force to a linear vibration system. It is known that the total general response of a system with an external force is the sum (superposition) of the steady state and particular (homogeneous) solutions. Because of this, the general response has a typical beating form (Figure 6.27a). The system natural frequency is equal to 1 radian per second, and the frequency of excitation is only increasing over time, so, again, we can choose the cutoff frequency of the lowpass filtering equal to 0.2.

Applying the proposed decomposition method, we will receive two main terms of motion. The first steady-state term (Figure 6.27b), separated from the solution by means of the HVD method, does not decay over time, while the second term, the transient component (free vibration), does decay (Figure 6.27c). After the free vibration part of the solution is damped, the system will oscillate according to excitation as long as the driving force is applied. The IF of the transient component (Figure 6.28a, dashed line) remains constant – close to the resonance frequency $1/2\pi \approx 0.16$ Hz, while the IF of the transient solution frequency increases at a linear rate (Figure 6.28a, bold line). Figure 6.28b also shows a pure exponential type of decay typical of a free vibration in linear systems, whereas the steady-state component has a more

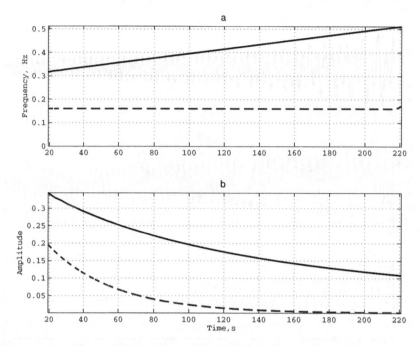

Figure 6.28   The IF (a) and envelope (b) of a nonstationary vibration solution: the steady state component (–), the transient component (- -)

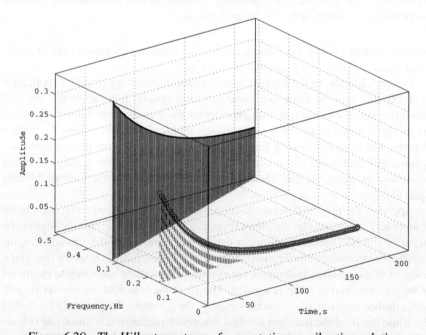

Figure 6.29   The Hilbert spectrum of a nonstationary vibration solution

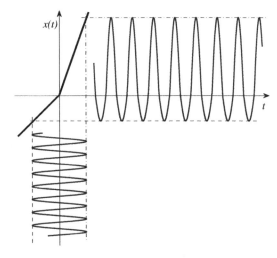

Figure 6.30   The asymmetric transformation of the signal amplitude

complicated envelope, depending on the frequency response function. Figure 6.29 includes the corresponding 3D plot of the IF and the envelope of every component. As can be seen in the Hilbert spectrum, where all decomposed waveform components are plotted in the time--frequency–amplitude domain, the composition in this case consists of two functions.

6.8.5   Asymmetric signal analysis

Sometimes a vibration signal consists of two independent separate parts: an upward motion associated with one function of the positive signal values and a downward motion associated with another function of the negative values:

$$x(t) = \begin{cases} x_1(t), \text{ if } x > 0 \\ x_2(t), \text{ if } x \leq 0 \end{cases} \tag{6.12}$$

The different positive and negative branches combined together produce an *asymmetric signal*. Each signal sign changing from positive to negative, or the reverse, switches the vibration structure, which will include the first or the second asymmetrical branch characteristics correspondingly. For example, a pure harmonic after its non-inertial and nonlinear transformation will form an asymmetric kind of a signal with different corresponding amplitude and frequency features. An illustration of an asymmetrical bi-linear amplitude transformation of the harmonics is shown in Figure 6.30.

The resultant signal combines independent positive and negative branches, so each part of the signal is determined only by its own instantaneous characteristics: $x_{1,2}(t) = A_{1,2}(t) \cos \left[\int \omega_{1,2}(t) dt \right]$ where $x_{1,2}(t)$ is the branch of the vibration signal, $A_{1,2}(t)$ is the *partial envelope* (the partial instantaneous amplitude), $\omega_{1,2}$ is the *partial frequency* (the partial IF) of the branch. One of the questions arising immediately due to the asymmetric representation is: How will the combined signal $x(t)$ be separated back into its constituent parts $x_1(t)$ and $x_2(t)$? Evidently, any standard approach fails in the case of the vibration signal asymmetry. The HT signal representation plays an important role in the asymmetric signal decomposition and leads directly to some practical results. Each signal branch is defined on its half-plane only, so practically it is enough to identify matching instantaneous characteristics of each signal branch.

For separating the signal positive and negative parts and for estimating the partial instantaneous characteristics we will use the already mentioned HVD method along with the congruent EOE approach (Sections 6.7 and 5.6) (Feldman, 2011). As a consequence an arbitrary aggregated signal will be built up from a slow-varying offset function and several alternate quasiharmonics with varying characteristics. In accordance with the HVD method all decomposed congruent quasiharmonic components will form the EOE function – according to their phase relations. The EOE aggregates all these component envelopes the following way:

$$A_{EOE}(t) = \sum_{l=1}^{N} A_l(t) \cos \phi_l(t), \qquad (6.13)$$

where $A_l(t)$ is the l component envelope, and $\phi_l(t)$ is the phase angle between the largest and the l components. Now we will form two EOE functions separately for the positive and negative asymmetric signal parts $A_{EOE}(t) = \begin{cases} A_p, & \text{if } x > 0 \\ A_n, & \text{if } x \leq 0 \end{cases}$. During a half of the period, when the largest harmonic of the signal is positive, the EOE appears as $A_p(t) = \sum_{l=1}^{N} A_l(t) \cos \phi_l(t)$; during the next half, when it is negative, the vibration continues as another EOE with the associated amplitude phase relations $A_n(t) = -\sum_{l=1}^{N} A_l(t) \cos \phi_l(t)$.

The IF of the asymmetric signal will match the frequency of the sequentially alternating positive and negative parts of the signal: $\omega(t) = \begin{cases} \omega_p(t), & \text{if } x > 0 \\ \omega_n(t), & \text{if } x \leq 0 \end{cases}$. As a first approximation, assume that the partial IF is a lowpass filtered IF of the corresponding part of the signal.

6.8.5.1   Estimation of asymmetric amplitudes

Consider an example of an asymmetric signal whose positive and negative parts have the same frequency but different amplitudes, while the positive partial amplitude increases linearly.

In Figure 6.31 we can observe the signal and that the asymmetry affects the instantaneous amplitude and frequency. In this case, the signal decomposed by the HVD method aggregates the first four elementary components shown in Figure 6.32. The first three components, including the increasing offset, are congruently in phase relative to maxima, but the fourth smallest component is out of phase. Therefore the partial positive envelope is equal to $A_p(t) = A_1(t) + A_2(t) + A_3(t) - A_4(t)$ and the partial negative envelope is equal to $A_n(t) = A_1(t) - A_2(t) - A_3(t) + A_4(t)$. Both calculated partial envelopes are shown in Figure 6.33 along with the initial asymmetric signal.

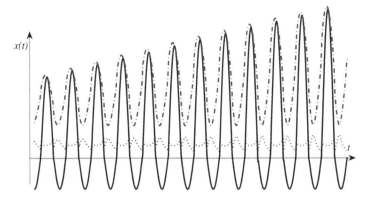

Figure 6.31   An asymmetric signal with a linearly increasing envelope: the signal (–), the envelope (---), the IF (. . .) (Feldman, ©2011 by Elsevier)

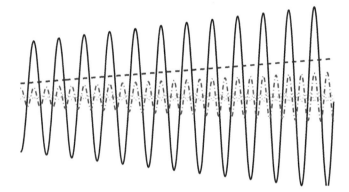

Figure 6.32   The decomposed components of an asymmetric signal with a linearly increasing envelope (Feldman, ©2011 by Elsevier)

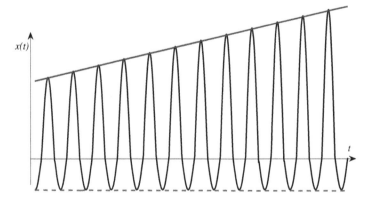

Figure 6.33   An asymmetric signal with a partial linearly increasing envelope: the signal (–), the positive partial envelope (–), the negative partial envelope (- - -) (Feldman, ©2011 by Elsevier)

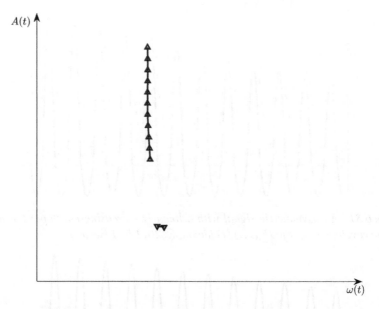

Figure 6.34 The envelope vs. the IF plot of an asymmetric signal with an linearly increasing envelope: the positive partial envelope ($\triangle$), the negative partial envelope ($\square$) (Feldman, ©2011 by Elsevier)

The partial frequencies for the positive and negative parts are the same, and the plot of partial envelopes vs. partial frequencies confirms an almost constant value for the partial frequency (Figure 6.34).

6.8.5.2 Estimation of asymmetric frequencies

Now consider another case of an asymmetric signal as a combination of two quasi-harmonics with the same amplitude but different partial frequencies ω_p, ω_n, while

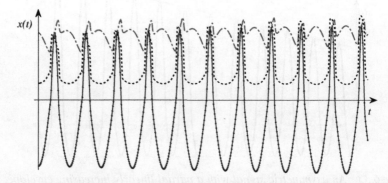

Figure 6.35 An asymmetric signal with a linear increasing frequency: the signal (−), the envelope (···), the IF (. . .)

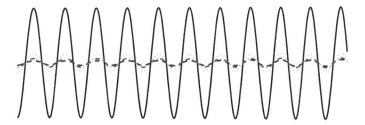

Figure 6.36   The decomposed components of an asymmetric signal with a linear increasing frequency

the frequency of the positive part linearly increases (Figure 6.35). Again, we use the HVD method to extract the signal components; the first four of which are presented in Figure 6.36. All the extracted components are congruently in phase, so each of the partial envelopes is equal just to the sum of the decomposed envelopes. The positive and negative partial envelopes both are shown with the initial signal in Figure 6.37.

The difference between the frequencies of the asymmetric parts of the signal leads to dissimilarity of the partial frequencies estimated after lowpass filtering of the instantaneous frequency segments (Figure 6.38).

6.8.5.3   Combined asymmetric amplitude and frequency signal

The identification method also operates well in the case of an asymmetrical amplitude and frequency combination. Consider the case of an initial asymmetric signal that consists of a linearly decreasing positive amplitude and a linearly increasing positive frequency (Figure 6.39).

As can be seen, the calculated signal envelope and the IF have a rather complicated form; nevertheless, they do not explain the asymmetry of the signal thus hiding the important information that is conveyed in this signal.

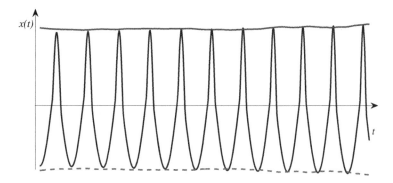

Figure 6.37   An asymmetric signal with a linear increasing frequency: the signal (–), the positive envelope (–), the negative envelope (- - -)

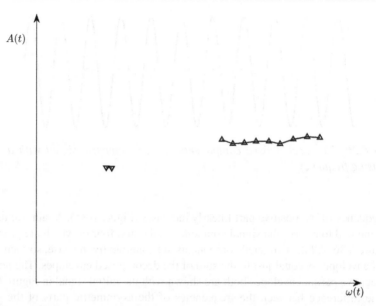

*Figure 6.38 The envelope vs. the IF plot of a asymmetric signal with a lin-
ear increasing frequency: the positive partial envelope (△), the negative partial
envelope (□)*

Again, the main idea of the method is to take an asymmetrical signal and divide
it into quasiharmonic components. The results of the signal decomposition obtained
according to the HVD method are illustrated in Figure 6.40, where the first four
estimated components are congruently in phase only for the positive part of the
largest harmonic. For the negative part the even harmonics are in phase, but the
odd harmonics are out of phase, so the positive partial envelope is $A_p(t) = A_1(t) +$

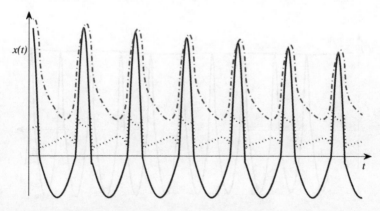

*Figure 6.39   An asymmetric signal with a linear decreasing envelope and increasing
frequency: the signal (–), the envelope (·-·), the IF (. . .)*

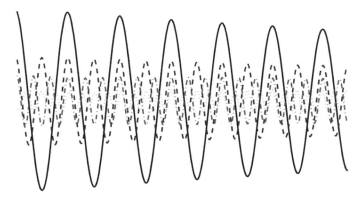

Figure 6.40 The decomposed components of an asymmetric signal with a linear decreasing envelope and increasing frequency

$A_2(t) + A_3(t) + A_4(t)$ and the negative envelope is $A_n(t) = A_1(t) - A_2(t) + A_3(t) - A_4(t)$ (Figure 6.41). The identified instantaneous characteristics are shown in Figure 6.42 and include the positive and the negative partial envelopes vs. their instantaneous frequencies.

As can be seen in Figure 6.42, the asymmetric signal separation behaves adequately. The proposed method seeks to overcome the limitations related to decomposing the asymmetric signal in a physically meaningful way. Having applied the method we obtained two groups of frequencies and envelopes associated with the positive and negative portions of the asymmetric signal. In other words, the initial signal is split into two portions separated for the positive and negative values. Use of

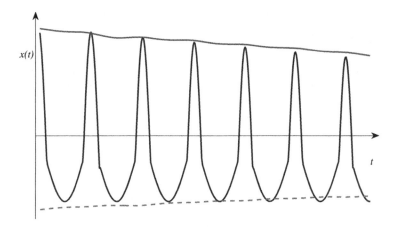

Figure 6.41 An asymmetric signal with a linear decreasing partial envelope and increasing frequency: the signal (–), the positive envelope (–), the negative envelope (--).

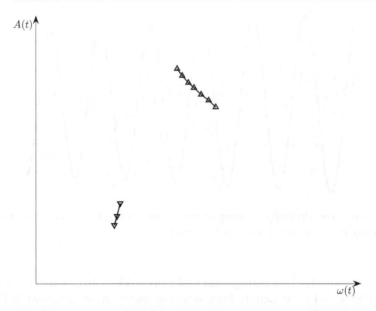

Figure 6.42   The envelope vs. the IF plot of an asymmetric signal with a linear decreasing envelope and increasing frequency: the positive partial envelope ($\triangle$), the negative partial envelope ($\square$)

the developed congruent technique could result in a more precise estimation of both the amplitude and the frequency of asymmetric vibration signals.

6.9   Comparison of the Hilbert transform decomposition methods

The EMD technique, an original technique first introduced by Huang *et al.* (1998), adaptively decomposes a signal into its simplest intrinsic oscillatory modes (components). The EMD method is based on a sifting iteration approach and on a spline algorithm, which constructs upper and lower envelopes that are fitted to the local maxima of the initial wideband signal.

Attempts to improve the EMD method were described in a large number of subsequent publications. For example, it was recommended that the original sifting approach should be replaced by a specialized nonlinear diffusion process for estimating the mean envelope, or replaced by an optimization-based approximation technique. Other attempts were devoted to developing alternate decomposition methods, like the iteration HT for modulated signals, which is based on the averaging or lowpass filtering of the modulated signal envelope (Gianfelici *et al.*, 2007). It is clear that averaging the envelope of a composition, even of two simple sinusoids, produces only an amplitude summation, but not the pure amplitude of any of the sinusoids. This means that the iteration HT cannot be applied to the general case of a multicomponent composition of non-modulated harmonics.

A different original method – the Hilbert Vibration Decomposition (HVD), which was developed by Feldman (2006) – is dedicated to the same problem of adaptive decomposition of nonstationary wideband vibration into certain basic monocomponents (atoms). As the HVD method is based on a lowpass filtering of the IF and does not involve spline algorithms, the filtered IF of each component will therefore be positive. Analytically, this lowpass filtered IF corresponds to the first term in the Blaschke polynomial approximation of the IF (Qian, 2006).

These two successful decomposition methods – namely, the local EMD and the global HVD, – have been theoretically analyzed and compared, and their common properties and differences are described in Feldman (2008b). Still, a detailed analysis of possible potentials and inherent shortages of these HT decompositions requires a further investigation of the EMD and HVD methods. Mapping the strength and recognizing the available resources of the HT decompositions will help us to compile and arrange a successful utilization of the methods. In addition to the Wavelets transform analysis, the HT decomposition method is a powerful approach that can solve rather complicated problems in various areas, including nonlinear and nonstationary mechanics and acoustics.

6.10   Common properties of the Hilbert transform decompositions

Two mentioned HT methods are dedicated to the same problem of decomposing nonstationary wideband vibration. The EMD is based on the spline fitting of the local extrema and the HVD method is based on averaging the global IF. We will try to analyze and compare the above-mentioned methods by investigating and understanding their general principles and limitations, but will not discuss concrete signal-processing procedures and algorithms. Both the local EMD and the global HVD signal decomposition methods are based on the assumptions that the underlying signal is formed by a superposition of quasiharmonic functions with or without an aperiodic slow-varying DC offset, and that the envelopes and the IF of each vibration component differ. These methods allow us to replace (provided that the replacement is possible at all) a nonstationary complicated waveform function with the composition of a small number of other functions that are simpler and more suitable for further computations or analytical transformations.

Both decomposition methods are nonparametric and adaptive because they deal with an *a priori* unknown nonstationary signal and do not require a model for the component representation. The EMD and HVD methods apply an iterative algorithm for a sequential extraction of components according to the sifting approach (Huang *et al.*, 1998). For the multicomponent signal with well-separated component frequencies, both the local and global methods produce good similar results by extracting the simplest component at each iteration step. With the HT decomposition approaches we can analyze amplitudes and frequencies of a large class of non-sinusoids but wavelike functions.

6.11   The differences between the Hilbert transform decompositions

The EMD method estimates both the local IF (via zero crossing) and the local envelope independently. By contrast, the HVD method primarily estimates the IF of the multicomponent signal. Then, the corresponding envelope is calculated on the basis of a lowpass filtered IF via synchronous detection. As a result, the EMD method is able to decompose most wideband signals with far separated frequencies. The HVD method, in addition to the wideband, makes a decomposition of the narrowband multicomponent signals also possible.

The local EMD method can detect a sharp envelope and/or frequency variations more accurately, whereas the global HVD method smears out the sharp envelope and frequency jumps and crossings. For the decomposition of multicomponent signals of type II or III (Section 5.3) with fast-crossing instantaneous parameters, a combined application of the known HT methods could also be recommended.

The local EMD method successfully operates with any data length, including a very short data, while the global HVD requires rather long data records. The HVD method has an additional feature: it detects signal components with desirable or specified frequencies – for example, those with only odd high harmonics of the main vibration component. The HVD method is extremely appealing in its simplicity, and can be recommended for many areas of signal processing, real-time digital signal processing included.

6.12   Amplitude—frequency resolution of HT decompositions

The decomposition methods that have been considered are both dedicated to discriminating a compound signal into a composition of simpler signals that can then be more easily modeled and investigated. A key element of a vibration decomposition is its ability to distinguish amplitudes and frequencies of the signal in time. It is clear that, according to the Heisenberg uncertainty principle, localization in both time and frequency is limited. A signal decomposition method for vibration analysis must be precise enough to primarily resolve adjacent frequencies and identify close oscillating components. The frequency resolution is the minimum difference in frequency between two harmonics that allows them to be resolved in a composition. The frequency resolution is one of the important indicators of the potentials of vibration decompositions.

6.12.1   The EMD method

As shown on Figure 6.9, the EMD is able to distinguish two harmonics in a composition only when their frequencies differ essentially. The limiting boundary of the EMD separation of two closely spaced harmonics is defined by the hyperbola-like form (Rilling and Flandrin, 2008; Feldman, 2009b):

$$(A_2/A_1)_{\text{boundary(bottom)}} \leq (\omega_2/\omega_1)^{-2} \tag{6.14}$$

where A_2/A_1 is the amplitude ratio and ω_1/ω_2 is the frequency ratio of the harmonics.

Depending on the amplitude and frequency ratio, the limiting boundary determines the region (3) where the EMD is able to separate the harmonics and the region (1) where the EMD cannot separate two tones (Figure 6.9). That is, the more the frequencies are spaced apart, the less is the amplitude ratio of two harmonics suitable for EMD separation. For example, a second harmonic with a tripled or lower frequency $f_2 \leq 3f_1$ and a small amplitude less than $A_2 \leq 0.11A_1$ cannot be separated by the EMD. This means that the EMD does not perform well for smaller amplitudes of the second harmonic and cannot distinguish frequencies that are close together. For example, if frequencies lie within an octave of each other $f_2 \leq 2f_1$ and their amplitudes differ by less than one-quarter $A_2 \leq 0.25A_1$, application of the EMD method is unable to separate these two components.

6.12.2   The HVD method

The HVD method controls the frequency resolution by applying a lowpass filter (see Section 6.7.1). Therefore, the IF of decomposed signal components will have its energy concentrated over a relatively small region in the frequency domain. The decomposed components can be separated only when their frequencies differ by more than the cutoff frequency of the filter and their amplitudes are not equal. Therefore, for high-frequency resolution the cutoff frequency of the lowpass filter should be as small as possible. Typically, the smallest cutoff frequency value of a stable and precise lowpass filter is $f_{\min} \geq 0.02Fs$, where Fs is the sampling frequency. For a harmonic with a frequency f_i that, for example, is sampled with 20 points per period ($f_i = 0.05F_s$), the filter can produce the next higher distinguishing frequency equal to $f_{i+1} = (0.05 + 0.02)Fs = 0.07Fs$. This means that the frequency components will differ by $0.07/0.05 = 1.4$, and several frequency components (0.05, 0.07, 0.09) lying in the same octave can be separated. Therefore the HVD method has a better frequency resolution capacity and can resolve much closer frequencies than the EMD. In fact, the data should be sufficiently long to detect not only low frequencies, but also small differences between frequencies that form a slow-beating effect in the signal.

6.13   Limiting number of valued oscillating components

6.13.1   The EMD method

The existing limit value of the EMD amplitude–frequency resolution (6.14) determines the maximum possible number of valued oscillating components that can be extracted from a wideband signal composition. To extract valued components we must sample for at least one complete cycle of every frequency. The lower observed

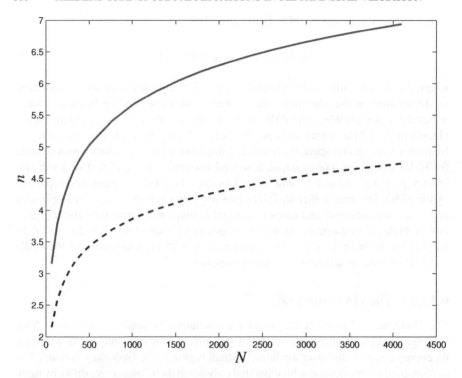

Figure 6.43   The largest number of valued oscillating components of the EMD: the frequency ratio is equal to 3 (–), the frequency ratio is equal to 5 (- -)

frequency that is sampled for a sufficient time to be detected, contains the total number of data N. Every next valued oscillating component with a higher frequency (6.14) will have fewer points, up to the highest frequency which, according to Nyquist theorem, will have not less than two samples. These neighbor components will be spaced far apart, with their frequency value (ω_{i+1}/ω_i) dictated by the EMD rough frequency resolution. As a result, the largest possible number of valued oscillating components n can be written in the following form

$$n < \log_{\omega_{i+1}/\omega_i} 0.5N, \qquad (6.15)$$

where n is a maximum number of valued oscillating components, N is the total number of samples, and ω_{i+1}/ω_i is the frequency ratio of the harmonics. An example of the limiting number of valued oscillating components is shown in Figure 6.43.

As can be seen in the figure, the case of a frequency ratio equal to $\omega_{i+1}/\omega_i = 3$ corresponds to the estimated largest number of valued components for $N \leq 1000$ samples not exceeding five harmonics. In the case of a frequency ratio equal to $\omega_{i+1}/\omega_i = 5$, the estimated largest number of valued components will not exceed three harmonics. In other words, the EMD is able to extract a rather small number of valued oscillating components with respect to the corresponding amplitude ratio and the data length.

The frequency resolution ability of the EMD can also be described in terms of the spectrum slope of a signal composition. Thus, doubling the frequency ratio corresponds to an octave, and the permitted minimum amplitude ratio corresponds to the slope of the spectrum. The limiting value of the changing amplitude with respect to the frequency takes the form, standard for engineers:

$$S_{dB} \leq 20 \log_{10} \frac{A_i}{A_0} = 20 \log_{10} 3 \approx \pm 10 \, \text{dB per octave.} \qquad (6.16)$$

The obtained spectrum slope ± 10 dB per octave defines a maximum gain change of 10 dB for each twofold increase or decrease in frequency. It is a low slope spectrum peculiar to a set of high harmonics with large amplitudes that differ not more than 10 dB for each doubling or halving of the frequency.

6.13.2   The HVD method

The HVD produces only a limited number of harmonics. But here the number of obtainable decomposed components depends on their frequency relation and the frequency resolution value. A sequence of separated frequencies represents an arithmetic progression such that the difference of any two successive frequencies is the value of the HVD frequency resolution. For example, in the case of the maximum frequency $f_{\max} \leq 0.16Fs$, the minimum frequency $f_{\min} \geq 0.02Fs$, and the best available frequency resolution $\Delta f \geq 0.02Fs$, where Fs is the sampling frequency, a limiting number of valued oscillating components will be not more than seven components:

$$n = 1 + \frac{f_{\max} - f_{\min}}{\Delta f} \leq 1 + \frac{0.16F_s - 0.02F_s}{0.02F_s} \leq 7 \qquad (6.17)$$

Therefore, the HVD is able to extract relatively more components than the EMD, but it is still a rather small finite number of valued oscillating components.

6.14   Decompositions of typical nonstationary vibration signals

Generally, a measured signal can be represented by a composition (sum) of a finite number of monocomponent signals: $x(t) = \sum A_l(t) \cos \left(\int \omega_l(t) dt \right)$, where $A_l(t)$ is an instantaneous amplitude and $\omega_l(t)$ is the IF of the l-component. In other words, the signal consists of a finite number monocomponents, where each has a constant or a slowly varying amplitude $A_l(t)$, a frequency $\omega_l(t)$, and an instantaneous phase $\int \omega_l(t) dt$. One of the most important results of the decomposition is an ability to preserve the phase content of the signal by constructing every initial component in the time domain and preserving all of its actual phase relations. The obtained combination of simplest components with the time and phase relations can provide us with some insight into a nonstationary vibration signal. Moreover, an improvement can be achieved in the analysis of the nonlinear dynamic system; but that cannot be said about other methods. The individuality of the simplest components inside the vibration composition allows the most effective decomposition method to be chosen.

6.14.1   Examples of nonstationarity vibration signals

6.14.1.1   Aperiodic DC offset plus oscillation

The DC offset can be easily removed from the vibration composition by different methods, including the well-known HT decomposition methods (Huang *et al.*, 1998; Feldman, 2006) (see Section 6.2.1). In the case of a short length data, the EMD method is more suitable for such a nonstationary offset extraction. The HVD method, which requires long length data, is also suitable for the offset extraction. It should be mentioned here, however, that only one single valuable aperiodic component can be decomposed by any method. The HT decompositions are unable to receive several slow-varying DC offset components, which means, for example, that the aperiodic DC offset cannot be further divided into increasing or decreasing components.

6.14.1.2   Sharp stepping-like variations in DC or in oscillating components

This case corresponds to well-separated but rapidly changing envelopes and IF trajectories, when the instantaneous characteristics as fast-varying functions can jump in time. In other words, such a multicomponent signal is composed of sequential segments with a step-varying DC, an envelope, and an IF. The EMD as the local method is more suitable for extracting sharp variations – for example, for detecting and synchronizing instant segments. The HVD in this case will extract only smoothed variations of instantaneous parameters.

6.14.1.3   Sum of quasi (almost) periodic oscillating-like harmonics

In the case of well-separated (non-crossing) and smoothed envelopes and IF trajectories, the energy and frequency of every component are well concentrated (localized), and the components do not overlap. These simple components resemble nonstationary modes like Fourier modes in the Fourier series, where every component can have a varying but non-crossing envelope and an IF. The EMD method allows separation of up to 3–4 harmonics with large amplitudes. The HVD allows separation of up to 5–6 harmonics with both large and small amplitudes.

6.14.1.4   Close-spaced and crossing-frequency amplitude components

In this case, the envelope and/or the IF trajectories have single or mutual crossings between each other. From a decomposition point of view, this is the most complicated type of multicomponent signal. The EMD method does not separate these close-spaced and crossing frequency components. However, the HVD method, which has relatively much better frequency resolution capacity, does decompose these nonstationary harmonics. An example of HVD separation of two crossing-frequency

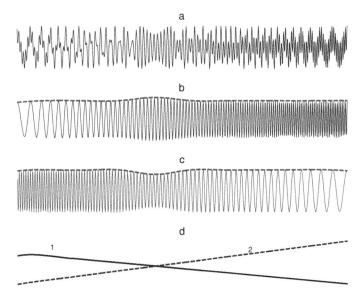

Figure 6.44   The HVD decomposition of two frequency- crossing components: the initial composition (a), the first decomposed component with its envelope (b), the second decomposed component with its envelope (c), the IF of the both components (d)

sweeping components is shown in Figure 6.44. In general, a decomposition result is not unique and is application-dependent.

6.14.1.5   Slow frequency modulated signal

A quasi-monoharmonic signal with a slow modulated frequency looks and acts like a harmonic during every short time interval, since at every moment the signal behaves as a single simplest harmonic when no further HT separation or decomposition is available. The EMD method results in a number of IMFs. Unfortunately, the decomposition of frequency modulated quasiharmonic signal does not always occur in the same single IMF. It is observed that the modulated signal arrives in different IMFs as broken time segments. Therefore, to get a physically meaningful decomposed frequency modulated signal we need to merge several IMFs together. The HVD method, on the contrary, results in a single frequency-varying component.

6.14.1.6   Amplitude modulated signal

An AM signal is generated by modulating the amplitude of a carrier signal, and it is composed of two terms: an oscillating carrier wave plus a wave that is a product of two harmonic-like terms. Actually, an AM signal is composed of three sinusoidal waveforms: a carrier wave, and lower and upper close-spaced sideband components. The EMD method does not separate these close-spaced components. However, in

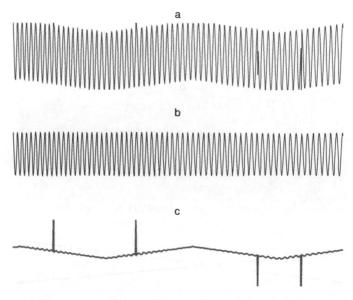

Figure 6.45   Extraction of the sweeping oscillations and exhibition of remaining impulses: the initial composition (a), the extracted sweeping oscillation (b), the de-noising impulses with a slow triangle component (c)

the case of amplitude modulation, the HVD method does allow separation of AM harmonics (see Sections 6.8.1–6.8.2).

6.14.1.7   Sum of impulses (sharp stepping-like varying components) and quasi (almost) periodic oscillations

The HT decomposition methods are not effective for the direct extraction of impulses in the signal. Nevertheless, the existing dominant interferences in the form of quasi (almost) periodic oscillations can be easily extracted from the composition. The remaining result, in the form of pure impulses, will be exhibited more informatively. In this case, instead of extracting impulses embedded in the noise, we will extract and remove the quasiharmonic noise and enhance the hidden impulse components. Such a HT-based filtering technique for structural noise removing (de-noising, noise cancelation, suppressing noise) offers excellent performance even when the impulse signal-to-noise ratio is very low (Figure 6.45).

6.14.1.8   Sharp (stepping-like) varying components

In the case of a complicated, sharp time-varying waveform (triangle, square, etc.), sometimes the purpose of a signal decomposition is not to receive a true frequency distribution, but rather to represent the complicated waveform in the form of the initial reduced time domain. The HVD method can further decompose the complicated waveform by spreading it out into the main quasiharmonic and high-frequency wave-like components. The HVD method can also perceive it only as a single sharp varying offset component without decomposition. The ability to decompose or not depends on the relation between the main component frequency and the cutoff frequency of the lowpass filter. If the main component frequency is less than the filter cutoff frequency, then the HVD perceives such a slow waveform as a single varying offset component. Otherwise, the method spreads it out into the main quasiharmonic and the high-frequency wave-like components.

The total number of obtained components – equal to the number of required iterations necessary to provide a good approximation of a signal composition – depends on how rapidly the initial waveform changes. As an example, Figure 6.23 (dashed line) depicts the signal of a nonstationary square wave function with a varying amplitude and period.

The first five decomposed component terms are plotted as a sum with the initial square wave in Figure 6.23 (solid line). As can be seen, it only requires the first five components to describe the nonstationary square wave in the time domain with a high degree of accuracy. This example illustrates the fact that the HVD method enables the construction of a perfect match between a complicated nonstationary signal and a composition of small number of time-varying elementary oscillating components. Contrary to that, the EMD method does not decompose the described sharp time-varying waveform.

6.14.1.9 Random signal

The EMD is able to filter a random signal as a bandpass filter bank, decomposing white noise into the IMFs whose frequency spectrum comprises an octave (Flandrin, Rilling and Goncalves, 2004). Every obtained IMF is the simplest fundamental component of a signal composition. As the simplest component, it cannot be further decomposed by the same EMD. An arbitrary data set can be reduced into IMF components only once. Applying the EMD to any IMF for a second time will produce nothing more than the same unchanged single IMF. Unlike the EMD, the HVD method does not decompose random signals.

All the considered types of nonstationary signal and the obtainable results of the HT decompositions have been summarized in Table 6.1.

6.15 Main results and recommendations

An arbitrary multicomponent nonstationary signal as a sum of quasiharmonic components with an aperiodic offset can be conventionally separated into elementary monocomponents by both the EMD and the HVD methods. Both methods successfully separate different frequency quasiharmonics and a single slow aperiodic component at each iteration.

There are at least two important limitations that should be kept in mind when applying the EMD and the HVD methods. The first limitation is the low-frequency resolution of the EMD method when the frequency range of the next IMF differs by more than one octave from the previous IMF. This is why the EMD method cannot

Table 6.1   Application of the HT decompositions for typical vibration signals

| Type of nonstationary signal composition | Obtainable results and recommendations | |
	EMD	HVD
Aperiodic DC offset plus oscillation	Extraction of DC offset for small and large length data	Extraction of DC offset for large length data
Sharp stepping-like variations in DC or oscillating components	Extraction of sharp variation	Extraction of smoothed variation
Sum of quasi (almost) periodic oscillating-like harmonics	Separation up to 3–4 harmonics with large amplitudes	Separation of up to 5–6 harmonics with large and small amplitudes
Close-spaced and crossing frequency components	No separation of harmonics with small amplitudes	Successful separation of harmonics
Slow frequency modulated signal	No separation of harmonics	No separation of harmonics
Amplitude modulated signal	No separation of harmonics	Successful separation of harmonics
Impulses (sharp stepping-like varying components) with quasi (almost) periodic oscillations	Extraction and removal of quasiperiodic oscillations and exhibition of remaining impulses	Extraction and removal of quasiperiodic oscillations and exhibition of remained impulses
Sharp stepping-like varying signal	No decomposition	Decomposition into quasiperiodic oscillations
Random signal	Bandpass filter bank filtration (decomposition)	No filtration (decomposition)

separate closely spaced or frequency-crossing harmonics. The HVD method has a better frequency resolution of even less than half of an octave, resulting in its capacity to decompose frequency-crossing nonstationary harmonics.

The second limitation is the finite small number of separated valued components, which does not exceed 3–4 components for the EMD and 5–6 components for the HVD. These limitations of the decompositions are often ignored in many vibration applications. Each decomposition method allows extraction of only a single slow-varying aperiodic offset, which displaces the initial composition by the aperiodic DC component.

Sometimes the extracted large quasiharmonic components are of no interest, but the remaining components, such as small impulses or sudden signal changes, are of interest for investigation. Thus, by removing large parts of the signal we are able to detect the very small remaining impulse components, even when the impulse signal-to-noise ratio is very low.

The decomposition methods are recommended for extracting a small number of nonstationary components with varying and *a priori* unknown instant parameters. The decomposition methods are dedicated primarily to decompose quasi and almost periodic oscillating-like signals. Typical examples of such signals are a time-varying waveform from a rotor startup or shutdown vibration (Antonino-Daviu *et al.*, 2007), a composition of steady-state and transient motions (Feldman, 2006), a vibration of nonlinear systems with superharmonics for precise identification (Feldman, 2007a), and so on. The decomposition methods are not very effective for the separation of other motion types, such as random (wideband noise), impulse, non-oscillating (aperiodic) vibration signals.

6.16   Conclusions

In this chapter, we considered some general properties of a multicomponent signal as a composition of narrowband nonstationary components with time-varying amplitude and frequency. The narrowband component is defined as a signal that always has a positive IF and envelope. The multicomponent signal allows us to extract a narrowband component from the composition, resulting in decreasing the spectrum bandwidth of the remainder of the signal.

A new and extremely simple Hilbert Vibration Decomposition method has been developed for vibration separation using the HT. Estimation of the varying frequency of the largest energy vibration component is effected by the lowpass filtration of the instantaneous frequency of the vibration. Synchronous envelope demodulation is performed by multiplying the composition by a sine and Hilbert projection wave, which are phase locked to the current component. This allows us to treat the nonstationary vibration composition as an aggregation of the synchronous components. The key factor of a precise decomposition is to use appropriate methods to extract the IF and envelope of the initial vibration composition.

A nonstationary example of the separation of a transient and a forced vibration regime was described; and an example of the decomposition of a time-varying vibration generated by a dynamical system obeying a nonlinear equation was shown.

Two signal decomposition methods (the EMD and the HVD) were analyzed and compared. It was analytically explained why the EMD generates the largest energy component by spline fitting and averaging of the local extrema. It was also explained why frequency resolution capabilities of the EMD method could be problematic. The decomposition methods were analyzed and compared in general, without a detailed inspection of signal-processing procedures such as spline approximation or filtering. The local and global HT decomposition methods can be used in combination to identify the intrinsic components more sharply and effectively.

The decomposition methods are dedicated primarily to decomposition of quasi and almost periodic oscillating-like signals. Such oscillating types could be, for example, multicomponent nonstationary modulated vibrations similar to a rotor startup or shutdown vibration, or the motion of a nonlinear dynamic system. The HT decomposition methods are not effective for separating other types of motion, such as random, impulse, non-oscillating (aperiodic) signals or signals from linear and time-invariant systems.

7

Experience in the practice of signal analysis and industrial application

Signal processing covers a number of applications for extracting information from measured data. Among other methods, the Hilbert transform (HT) provides some unique information on the nature and level of the measured vibration signal (Randall, 1986). The information describing a signal is carried out mainly by the instantaneous amplitude, phase, and frequency (Figure 7.1). These functions instantly monitor any change in a physical phenomenon, source, and medium through which a signal is transmitted. The key to an accurate vibration characterization is to decompose complex time signals into functions of different characteristic time scales and extract their time-varying frequencies and amplitudes (Pai, 2007; Gendelman, Starosvetsky, and Feldman, 2008; Starosvetsky and Gendelman, 2008, 2010).

In addition to a direct extraction of the instantaneous amplitude, phase, and frequency from the vibration signal, the HT gets information about disturbances and distortions of the instantaneous signal attributes by utilizing a signal demodulation. Furthermore, it provides a decomposition of the complicated signal and constructs a frequency–time distribution of the energy of the motion as a Hilbert amplitude spectrum and smooth congruent envelopes. The HT is an effective tool for analyzing unsteady transient vibration signals. It can clearly indicate frequency–amplitude differences with time and extract some time–frequency characteristics that cannot be obtained by any other method (Guo and Peng, 2007; Huang and Shen, 2005).

Hilbert Transform Applications in Mechanical Vibration, First Edition. Michael Feldman.
© 2011 John Wiley & Sons, Ltd. Published 2011 by John Wiley & Sons, Ltd.

$$x(t) = A(t) \cos \phi(t) = \sum A_i(t) \cos \int \omega_i(t) dt$$

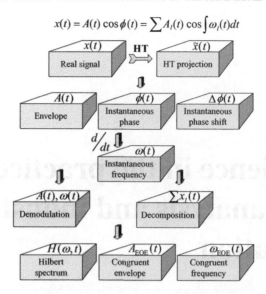

Figure 7.1   The HT procedures in signal processing (Feldman, ©2011 by Elsevier)

7.1   Structural health monitoring

7.1.1   The envelope and IF as a structure condition indicator

Vibro-acoustic modulation is by far the most widely exploited approach for monitoring various nonlinear symptoms from acoustical responses. The intensity of modulation is directly related to the severity of damage. Various parameters, based on the amplitude of the carrier frequency and modulation sidebands, have been used to describe this severity. A recent paper (Hu *et al.*, 2010) devoted to the demodulation of the envelope and the IF, explored the time-domain analysis of modulated acoustical responses. This investigation focused on the instantaneous characteristics of the response using the HT. The study showed that both modulations – that is, amplitude and frequency – are present in the acoustical responses when the aluminum plate is cracked. However, the intensity of amplitude modulations correlates far better with crack lengths than the intensity of frequency modulations. A nonlinear acoustic test was performed for an undamaged and cracked aluminum plate instrumented with surface-bonded, low-pro le piezoceramic transducers.

The concept of the IF as a potential candidate for a damage detection indicator was examined in Bernal and Gunes (2000). The rst step in the use of the EMD method was to decompose the signal into several monocomponent signals. This improves the likelihood of the IF concept being negative. In the case of a sudden severe damage, when the structure remains linear after the damage, the technique was capable of identifying the time and the extent of the damage. It was shown that, although the computations were made for noiseless conditions, the approach can give useful information in realistic conditions with noisy measurements. In the case of a hysteretic response, the IF was found to be inadequate as a robust indicator of modest nonlinearity; but in the case of severe nonlinear behavior, the IF showed some clear trends consistent with inelastic behavior. The IF was also recommended for the interpretation of a crack-induced rotor nonlinear response (Yang and Suh, 2004a).

In Salvino *et al.* (2005) the EMD was applied in structural health monitoring. Time–frequency features and instantaneous phase relationships were extracted and examined for changes that may occur due to damage. It was shown that the EMD and instantaneous phase detection approach, based on time–frequency analysis along with simple physics-based models, can be used to determine the presence and location of a structural damage.

The HT and AM–FM signal-modeling technique has been applied to a loud-speaker identi cation (Grimaldi and Cummins, 2008). In order to characterize the speaker, a single IF for a real-valued signal was constructed rst. An importance of the IF stems from the fact that speech is a nonstationary signal with spectral characteristics that vary with time.

The HT method was also effective for resolving the true physical features characteristic of a dynamic motion experiencing nonlinearity and undergoing bifurcation in a rotor-journal bearing model (Yang and Suh, 2004b). A model that was subjected to the breathing and slow growth of a transverse surface crack was developed. It was employed to demonstrate the effectiveness of the method in characterizing the inception and progression of various states of bifurcation. The fundamental notion of an IF de nes frequency as a phase temporal gradient, and thus provides a powerful mechanism to dissociate amplitude modulation and frequency modulation. The results of applying the IF to the characterization of bifurcation, and the evolution of instability, for a cracked rotor also indicate that the IF interprets nonlinear rotary responses with sound physical bases (Salvino et al., 2005).

7.1.2   Bearing diagnostics

Based upon the EMD and the Hilbert spectrum, a method for the fault diagnosis of roller bearings is proposed in Yu, Cheng, and Yang (2005). The envelope and the local Hilbert marginal spectrum are used to detect fault patterns and to monitor a roller bearing with outer-race faults or inner-race faults. The results show that the proposed method is superior to the traditional envelope spectrum method in extracting the fault characteristics of roller bearings.

An application of the EMD method to the vibration analysis of ball bearings is introduced in Du and Yang (2007), where the local mean of the extrema is computed by averaging successive extrema. Based on the improved EMD method, the vibration signals of ball bearings are analyzed in detail, and it is shown that the proposed method is superior to the discrete wavelet decomposition for the vibration analysis of ball bearings. For more about the HT and the structural health monitoring relations see, for example, Boller, Chang, and Fujino (2009).

An EMD method, improved to restrain the end effect in EMD, was applied to the fault diagnosis of large rotating machinery (Wu and Qu, 2008). The EMD provided an easier and clearer approach to the fault diagnosis by an investigation of the IMFs in the time domain. A radial rub between the rotor and stator of the machine is a serious malfunction that may lead to catastrophic failure. It normally involves several physical effects, such as friction, impacting, and nonlinear behaviors in the rotor-bearing system. For instance, it was shown that – due to a radial rub between the rotor and the stator – the shaft orbit suddenly changed its elliptical trajectory. Thus, an acute change of shaft orbit curvature can be noticed and considered as a feature of the radial rub. The observed amplitude modulation component from the compressor was caused by an abnormal excitation in a pipe.

7.1.3   Gears diagnosis

Data acquisition for the testing or analysis of rotating machinery usually involves measurements of the speeds of the rotating elements. It is sometimes the case, however, that space limitations make it dif cult to introduce speed measuring devices into the system. Measuring the speed of gears that are already in the system requires a minimum of space and system disturbance. However, the resolution of the measurement is limited by the number of teeth on the gear. A method presented in Wallace and Darlow (1988) uses HT techniques to recover the shaft oscillations of a frequency approaching that of the passing teeth. The HT can be a common and easily instrumented approach to monitoring shaft speed with magnetic transducers mounted near the teeth of a shaft-mounted gear. This work demonstrates that transient or higher frequency shaft vibrations can be recovered from this type of data. An example of the use of HT for signal demodulation is presented and compared to a time-domain approach for measuring transient speed variations. The HT technique has been shown to be an accurate method of recovering transient gear speeds when high-resolution devices cannot be attached to the shaft.

Two envelope analysis methods performing bearing diagnostics are suggested in Ho and Randall (2000). It has been found that analyzing the squared envelope can improve the signal to noise ratio in certain situations. If the ratio is greater than unity, there is an advantage in analyzing the squared envelope rather than the envelope itself, because the ratio increases as a result of the squaring operation. The modulation frequencies representing the bearing fault are always present additively in the envelope spectra, but they can be masked by discrete or random noise.

The EMD and Hilbert spectrum were applied to the vibration signal analysis for the fault diagnosis of a localized gearbox (Liu, Riemenschneider, and Xu, 2006). Vibration signals collected from an automobile gearbox with an incipient tooth crack were used in the investigation. The results show that the EMD algorithms and the Hilbert spectrum perform excellently. They are found to be more effective than the frequently used continuous wavelet transform in the detection of the vibration signatures. The effects of modulation and nonstationarity in vibration signals collected from the faulty gearbox present challenges for the extraction of fault features (Fan and Zuo, 2007). By applying demodulation and time–frequency analysis, the HT generates a feature indicator representing the real gearbox condition through an analysis of the vibration signal. Comparison studies show that the proposed method is more effective and does not require operators to have much diagnostic experience.

7.1.4   Motion trajectory analysis

Vibration motion measured by two sensors in two different directions forms a vibration trajectory, or an orbit, as a path of the body on a plain. The simplest form of motion trajectory is a shaft whirling that often accompanies rotation as some forces, like unbalance and surrounding uids exert a force on the rotating elements. Typically, two orthogonal sensors are positioned at a right angle to measure the shaft motion. The HT can be employed to decompose and separate the forward and backward whirl direction (Bucher *et al.*, 2004). Examining the whirling motion of a shaft, it is clear that in a forward whirl orbit the rst direction precedes the second by 90°, while in a backward whirl the rst lags the second direction response by 90°. This fact is true for every time moment of every frequency; therefore a nonstationary signal can be separated by using the HT to create two signals representing forward and backward whirl orbits, respectively. The HT approach illustrates the main advantage of using the forward/backward decomposition over the frequency and time domains (Bucher and Ewins, 1997; Bucher, 2011; Lee and Han, 1998). An instantaneous estimate of the forward and backward components containing all the frequencies of the original signals is obtained. A possible application of this real-time decomposition is an online diagnostics.

7.2   Standing and traveling wave separation

The HT may be of practical use for the real-time kinematic separation of nonstationary traveling and standing waves. Whenever these two wave components are traveling through the same medium at the same time, they pass through each other without being disturbed. According to the principle of superposition, the transverse vibration of the medium at any point in space or time is simply the sum of the individual wave components, namely, the traveling and standing components. A traveling wave moves from one place to another, whereas a standing wave appears to stand still, vibrating in situ. Consider a pure traveling wave of the form (Feeny, 2008):

$$W_{tr}(x, t) = A_{tr} \cos(kx) \cos(\omega t) \pm A_{tr} \sin(kx) \sin(\omega t). \tag{7.1}$$

where $A_{tr} \cos(kx) \cos(\omega t)$ is a component of the cosine wave, and $A_{Tr} \sin(kx) \sin(\omega t)$ is a component of the quadrature wave. The sign of a quadrature wave component de nes the right, or left, direction of the propagation of a traveling wave.

Another pure type of solution of the wave equation is a standing wave that oscillates in place, but does not travel (translate in space). Such a wave, also called a mode shape, has the form: $W_{st}(x, t) = A_{st} \cos(kx + \delta_0) \cos(\omega t)$. Standing wave characteristics are locations with maximum displacement (antinodes) and locations with zero displacement (nodes). The locations where $\cos(kx + \delta_0) = 0$ are called nodes, whereas the places where $\cos(kx + \delta_0) = \pm 1$ are called antinodes. The initial phase angle δ_0 describes the relative disposition of the mode shape along the vibration body. Let us choose an original (an initial, or a starting) point for the standing wave ($x = 0$) at the antinode point where the standing wave has a maximum $kx + \delta_0 = 0$. In this case, the initial phase angle δ_0 is also equal to zero and the standing wave takes the simplest form:

$$W_{st} = A_{st} \cos(kx) \cos(\omega t) \tag{7.2}$$

In a general case, the two wave components described above form a signal composition: $W_\Sigma = W_{tr} + W_{st} = (A_{tr} + A_{st}) \cos(kx) \cos(\omega t) \pm A_{tr} \sin(kx) \sin(\omega t)$, that can be normalized to have a maximum lateral displacement of unity:

$$W_\Sigma = \cos(kx) \cos(\omega t) \pm \gamma \sin(kx) \sin(\omega t), \tag{7.3}$$

where $\gamma = A_{tr}/(A_{tr} + A_{st})$ is the traveling wave ratio, meaning a portion (scalar measure) of the amplitude of the traveling wave in the total amplitude of the composition. A traveling wave ratio always falls in the range [0, 1].

The separation of vibration waves into their traveling and standing components requires simultaneous spatial information, typically obtained from an array of two or more sensors, while one of the sensors is a reference (original, starting) point. Theoretically, the reference point can be taken at an arbitrary point on the oscillating body. By choosing the reference signal at the point nearest to the antinode, where the standing wave has its maximum (7.2), the standing wave takes the form $W_{st} = A_{ref} \cos(\varepsilon) \cos(kx) \cos(\omega t) \approx A_{st} \cos(kx) \cos(\omega t)$, where $A_{st} = A_{ref} \cos \varepsilon$ is the reference signal amplitude value and ε is the phase shift in the space between the real antinode and the reference point. The value of ε depends on the position of the sensor and on the total number of measurement points along the wave. The largest possible value of ε in the case of equally spaced points is equal to $\varepsilon_{max} = 0.5\pi/(1 + n/4)$, where n is the total number of points for a period of the standing wave. For example, for more than 16 points, the measured amplitude of the reference point will be only 5% less than the real amplitude at the antinode. In practice, this error in amplitude values is small and can be neglected.

The HT, being a linear operator, acts upon the sum of the individual wave components – namely, the traveling and standing components – independently. Thus, the composed vibration signal W_Σ (7.3) along the wave will have the following envelope and phase shift functions:

$$A(x)^2 = \gamma^2 \sin(kx)^2 + \cos(kx)^2; \quad \phi = \arctan\left[\gamma \sin(kx)/\cos(kx)\right], \tag{7.4}$$

where $A(x)$ is the measured amplitude and ϕ is the measured phase angle relative to the phase in the initial antinode point, where $x = 0$. The instantaneous relative phase shift in the case of two different nonstationary real signals, $x_0(t)$ and $x_i(t)$,

can be estimated as the instantaneous relative phase between them, according to the
HT (2.8): $\phi = \arctan \dfrac{x_0(t)\tilde{x}_i(t) - \tilde{x}_0(t)x_i(t)}{x_0(t)x_i(t) + \tilde{x}_0(t)\tilde{x}_i(t)}$, where $\tilde{x}(t)$ is the HT projection of the
measured real signal $x(t)$.

To separate a composed vibration wave into two components we need
to nd an unknown traveling wave ratio γ. The estimated envelope and the
phase shift of the measured vibration signals in two points allow this unknown
parameter to be calculated directly. Rewriting expressions (7.4) in the form
$\cos(kx)^2 = A^2 / [1 + \tan(\phi)^2]$, $\gamma^2 \sin(kx)^2 = A^2 / [1 + \cot(\phi)^2]$ and utilizing the
trigonometric identity $\cos(\varphi)^2 + \sin(\varphi)^2 = 1$ we will obtain the unknown traveling
wave ratio

$$\gamma^2 = \frac{\sin(\phi)^2}{A^{-2} - \cos(\phi)^2}. \tag{7.5}$$

Having identi ed γ, the wave decomposition can be formed. It is composed of a normalized traveling component $\gamma[\cos(kx)\cos(\omega t) + \sin(kx)\sin(\omega t)]$ and a normalized
standing component, in the initial antimode point: $(1 - \gamma)\cos(kx)\cos(\omega t)$. In a normalized form, γ can attain a maximum value of 1; in which case the standing-wave
component vanishes.

The traveling wave has a wavelength $L = 2\pi / kx_{peak}$ as the distance between
two neighbor peaks, oscillating with the frequency ω. Thus, the propagation speed
V_{tr} of the traveling wave – as a rate at which a given peak of the wave travels – has a
known form: $V_{Tr} = \omega L / 2\pi = \omega / kx_{max}$.

As an illustration to the proposed technique, we use a simulated example of
a composed vibrating string with 32 vibration points. A traveling wave ratio is
calculated directly and is equal to $\gamma = 0.35$. Figure 7.2 shows the traveling part
moving horizontally with the propagation speed V_{tr}. Figure 7.3 shows the standing
part of the wave which simply oscillates and does not travel to the right or to the left.

The suggested use of the HT for a direct kinematics separation of the traveling
and standing components may be a practical method for the real-time identi cation
and tuning of the traveling waves (Minikes *et al.*, 2005) and a damage visualization
(Ruzzene, 2007).

The application of the HT to the study of internal gravity waves can provide interesting results and answer questions still unsolved (Mercier, Garnier, and
Dauxois, 2008). Thanks to the analytical representation of the internal waves using
the HT, it is easy to obtain the envelope of a monochromatic internal wave and thus
quantify how it decreases through viscous dissipation. The HT is an excellent tool
for measuring the dissipation effects. The HT complex demodulation presented in
Mercier, Garnier, and Dauxois (2008) not only offers an analytical representation of
the wave eld with extraction of the envelope and the phase of the waves, but allows
a discrimination of different possible internal waves of one given frequency. The
experimental investigation of the attenuation, re ection, and diffraction of internal
plane waves generated using a new type of generator has provided answers to several
theoretical assumptions that had never been con rmed.

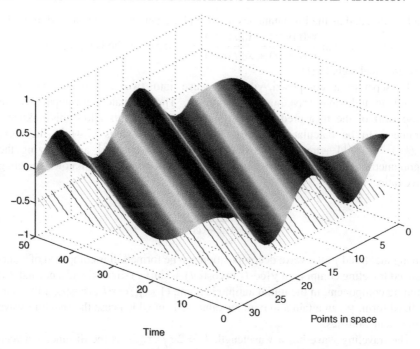

Figure 7.2   The traveling part of a wave (Feldman, ©2011 by Elsevier)

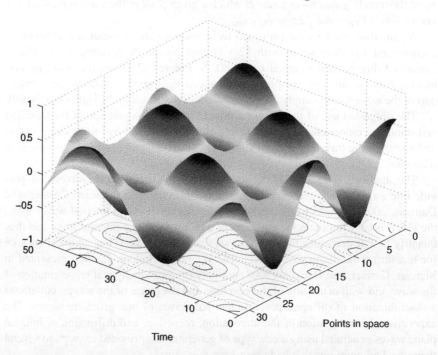

Figure 7.3   The standing part of a wave (Feldman, ©2011 by Elsevier)

7.3   Echo signal estimation

The HT is used to analyze an echo signal that appears at a time that overlaps with the initial signal itself. Such a HT approach is proposed to identify the arrival time of overlapping ultrasonic echoes in a time-of- ight diffraction aw detection (Chen *et al.*, 2005), for example. A nonlinear and nonstationary ultrasonic signal was decomposed using the EMD method to obtain intrinsic mode functions, which were used for a signal reconstruction. By applying the HT to the reconstructed signal, the arrival time of each echo can be clearly identi ed. The ef ciency and feasibility of this approach in enhancing the time resolution of an ultrasonic signal are validated by a simulation and an experiment, where 98% of aws in 12-, 10-, and 8-mm thick pipelines could be identi ed with an average accuracy of 0.2 mm. The blind area of the diffraction has been reduced to 2.5 mm under the surface. The frequency and angle of the probe pair have little in uence on the identi cation (Chen *et al.*, 2005).

7.4   Synchronization description

The rst key effort in the use of the HT to characterize the effect of a phase synchronization of weakly coupled self-sustained chaotic oscillators was made by Rosenblum (1993) and Rosenblum, Pikovsky, and Kurths (1996). These authors observed a synchronization phenomenon for coupled Rössler attractors, where phases were locked in a synchronous regime, while amplitudes varied chaotically and were almost uncorrelated. Coupling a chaotic oscillator with a hyperchaotic one, produced a new type of synchronization, where the frequencies were entrained, while the phase difference was unbounded. A relation between the phase synchronization and the properties of the Lyapunov spectrum was also studied.

The HT and the analytic signal representation are widely used in the investigation of the modern nonlinear dynamics of chaotic and stochastic systems (Anishchenko *et al.*, 2007).

7.5   Fatigue estimation

The HT approach can be used to count fatigue cycles in an arbitrary loading waveform. It processes a time history representing the random wideband loading condition to generate the number of cycle-counts with their corresponding amplitudes. The approach is general, accurate within any desirable degree, and amenable to modern signal processing (Kendig, 1997; Gravier *et al.*, 2001).

7.6   Multichannel vibration generation

The HT can be used not only for analysis, but also to construct the required signal (Manske, 1968; Yang, 1972). For a more detailed and general discussion see Liang, Chaudhuri, and Shinozuka (2007), which presents a simulation of one-dimensional, univariate, nonstationary stochastic processes by integrating Priestley's evolutionary spectral representation theory. Applying this simulation, sample functions can be generated with great computational ef ciency, while the HT gives more concentrated energy at certain frequencies. The simulated stochastic process is asymptotically Gaussian as the number of terms tends to in nity. A mean acceleration spectrum, obtained by averaging the spectra of generated time histories, are then presented and compared with the target spectrum to demonstrate the usefulness of this method (Liang, Chaudhuri, and Shinozuka, 2007).

7.7   Conclusions

In conclusion, it may be said that some general properties of a multicomponent signal were described. The signal was presented as a composition of narrowband nonstationary components with time-varying amplitudes and frequencies. A narrowband component is de ned as a signal with an always positive IF and envelope. The multicomponent signal permits a narrowband component to be extracted from the composition, which decreases the spectrum bandwidth of the remainder of the signal.

Two signal decomposition methods (the EMD and the HVD) were analyzed and compared. It was analytically explained why the EMD generates the largest energy component by spline tting and averaging of the local extrema. It was also explained why frequency resolution capabilities of the EMD method could be problematic. The decomposition methods were analyzed and compared in general, without detailed inspection of signal-processing procedures such as a spline approximation or ltering.

The decomposition methods are dedicated primarily to decompose quasi and almost periodic oscillating-like signals. Such oscillating types could be, for example, multicomponent nonstationary modulated vibrations similar to rotor startup or shutdown vibrations, or have a motion like a nonlinear dynamic system. The HT decomposition methods are not effective for the separation of other types of motion, such as random, impulse, nonoscillating (aperiodic) signals, or signals from linear and time-invariant systems.

The text in Part II has described the main results and some limitations of two HT decompositions – the local EMD and the global HVD methods. The clari ed issues regarding amplitude–frequency resolution will often give a better signal representation. The typical nonstationary signals that were considered – with applications and practical recommendations – constitute an attempt to adjust and more completely realize the potentials of HT signal processing, and, subsequently, to improve the nonstationary and nonlinear data analysis.

According to the knowledge of a large body of research, HT signal processing yields excellent results concerning the detection of different kinds of damage and is applicable for online structural health monitoring. Generally, the success of a signal-based approach depends very much on the initial physical knowledge of the user. It may yield valuable information about the type of damage, but the extent can only rarely be calculated and the results are of a qualitative nature. However, knowledge of the type of damage detected by the signal-based approach is a very valuable basis for an improved model-based diagnosis that will yield a quantitative estimate of the size and location of the damage. A purely model-based approach, on the other hand, can fail if no initial knowledge of the damage is available. In such a case a multitude of different types of damage would have to be considered, and it would be a dif cult (and time-consuming) task to distinguish damages that had similar effects on the measurements. True progress in the area of diagnostics can be made if signal- and model-based diagnoses are combined.

Part III

HILBERT TRANSFORM AND VIBRATION SYSTEMS

8

Vibration system characteristics

The physical phenomena in vibration systems can be studied either in the time or frequency domains involving different mathematical descriptions (Hammond, 1968). These descriptions are equivalent and are related by the Volterra series and the HT (Worden and Tomlinson, 2001). The choice of a domain is merely for mathematical convenience. The time (temporal) description of the process can be regarded as a "natural" description for non-stationary processes, while the spectral description is a measure of the power distribution over the frequency.

The HT and analytical signal can also be successfully applied directly to differential nonlinear equations of motion. Thus the analytic signal becomes a core of the method of separating the frequencies in nonlinear systems (Vainshtein and Vakman, 1983). The method separating the frequencies uses a nonlinear vibration equation, analytically considers its higher approximations, and derives a solution that contains different high-order superharmonics.

8.1 Kramers–Kronig relations

There is a number of quantities that may be used to characterize a linear vibration system. Among the most popular is a complex function of frequency which contains complete information describing the behavior of a linear system after its excitation with an arbitrary stimulus. A frequency response function (FRF) is often expressed using a complex notation that makes it possible to combine the modulus $|\text{FRF}(\omega)|$ and the phase $\phi(\omega)$ into one complex function $\text{FRF}(\omega) = \text{Re}(\omega) + i\,\text{Im}(\omega) = |\text{FRF}(\omega)|\,e^{i\phi(\omega)}$. The real part $\text{Re}(\omega)$ and the imaginary part $\text{Im}(\omega)$ of a complex frequency function are related by the HT, generally known among physicists as the Kronig–Kramers relation (King, 2009). For

Hilbert Transform Applications in Mechanical Vibration, First Edition. Michael Feldman.
© 2011 John Wiley & Sons, Ltd. Published 2011 by John Wiley & Sons, Ltd.

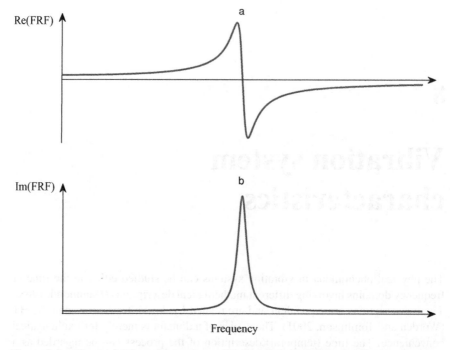

Figure 8.1   The real (a) and imaginary (b) parts of the FRF of a linear vibration system (Feldman, ©2011 by Elsevier)

a linear causal system the HT presents interesting mathematical properties of the Kramers–Kronig relation (MacDonald and Brachman, 1956; Pandey, 1996). The properties connect the real and imaginary parts of any complex analytic function in the upper half plane. We can associate the imaginary part of a complex frequency response function as the HT function of the real part. In other words, the real and imaginary parts of the transfer function form a Hilbert pair (Figure 8.1):

$$\mathrm{Re}\,(\omega) = H\,[\mathrm{Im}\,(\omega)]; \quad \mathrm{Im}\,(\omega) = -H\,[\mathrm{Re}\,(\omega)] \tag{8.1}$$

A relation between the modulus $|\mathrm{FRF}(\omega)| = \sqrt{\mathrm{Re}\,(\omega)^2 + \mathrm{Im}(\omega)^2}$ and the phase $\phi(\omega) = \arctan\,[\mathrm{Im}(\omega)/\mathrm{Re}\,(\omega)]$ of the FRF is also provided formally by the HT (2.1) (Figure 8.2): $\phi(\omega) = -H\,[\ln A(\omega)]$. A group delay (Figure 8.2b) is the phase   rst derivate of the FRF $\Delta t_{\mathrm{group}} = -d\,[\phi(\omega)]/d\omega$ and shows the time needed for signal components around the speci c frequency to pass through the system.

The imaginary part of a response function describes how a system dissipates energy since the response is out of phase with the driving force. These relations imply that observing the dissipative response of a system is suf cient to determine its inphase (reactive) response, and vice versa. The energy dissipation in a vibration system is a known effect of the change in the phase frequency function. The relationship between dissipation and attenuation is a direct consequence of causality and linearity. The occurrence of nonlinearity can manifest itself as a disability of many linear time-series analysis techniques to describe the dynamics accurately. Any departure

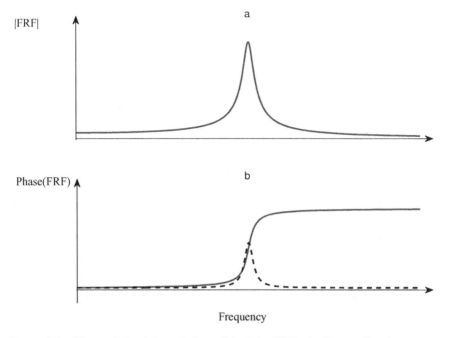

Figure 8.2   The modulus (a), and phase (b) of the FRF of a linear vibration system and group delay (– –) (Feldman, ©2011 by Elsevier)

from an initial linear frequency response function (that is, a distortion) can be attributed to nonlinear effects (Simon and Tomlinson, 1984; Tomlinson and Ahmed, 1987; Worden and Tomlinson, 2001). Any deviation from a linearity could be conceivably taken as a measure re ecting the degree of the nonlinearity; however, it will require some *a priori* knowledge of the system under study, that is, we will need to know how the distortion manifested itself. Thus, the HT in the frequency domain can be a diagnostic tool that discovers a nonlinearity on the basis of the measured FRF data. It merely exploits the fact that the FRF of a linear system is invariant under the Hilbert transformation.

Many of the standard signal-processing tools intended for the FRF assume that the underlying vibration system is linear, and are therefore unable to capture nonlinear relationships from the measured time-series data directly. The HT applied to the FRF function analysis allows us to detect the presence of nonlinearities. Note that dynamic systems with bifurcations or a chaotic behavior are not part of the considered class of nonlinear systems.

8.2   Detection of nonlinearities in frequency domain

Historically, most of the rst developments of HT applications in vibration systems (e.g., a nonlinearity detection) have been using a frequency domain HT analysis (Kerschen *et al.*, 2006). Possibly, it happened because nonlinear systems produce vibration response components as a group of frequencies other than a single frequency of the pure harmonic excitation. This spectral distortion in the frequency domain indicates the existence of a nonlinear effect and can be detected by many methods. For example, a nonlinear response can be expressed in terms of Volterra series expansion, higher-order FRFs, and higher-order spectra (Worden and Tomlinson, 2001). There exists a practice – in frequency domain methods – to inspect FRFs through a visualization of Nyquist plots which combine the gain and phase characteristics in a single complex plain. Since a Nyquist plot of each well-separated resonance mode of a linear vibration system shows a circle, a linear FRF is very easily recognized and the existence of a nonlinearity can be discovered by examining its abnormal behavior when a departure from the circle line indicates the presence of a nonlinearity or noisy measurement conditions (Goge *et al.*, 2005).

One more diagnostic tool for distortions indicative of a nonlinearity, introduced in Simon and Tomlinson (1984), is provided by a HT applied to the measured FRF data. It merely exploits the fact that the FRF of a linear system is invariant under a Hilbert transformation. Really, the HT is a linear operator, so for constants α, β and functions f and g it follows that $H\left[\alpha f(\omega) + \beta g(\omega)\right] = \alpha H\left[f(\omega)\right] + \beta H\left[g(\omega)\right]$ (King, 2009). The HT of a real function results in a real function, and similarly that of a complex function results in a complex function: $H\left[f(\omega) + ig(\omega)\right] = H\left[f(\omega)\right] + iH\left[g(\omega)\right]$. For the response function in the frequency domain $\mathrm{FRF} = \mathrm{Re}(\omega) + i\mathrm{Im}(\omega)$ a unique HT relationship exists between its real and imaginary parts (8.1). This allows us to write the following expression for the Hilbert transformed frequency response function: $H\left[\mathrm{FRF}(\omega)\right] = H\left[\mathrm{Re}(\omega)\right] + iH\left[\mathrm{Im}(\omega)\right] = -\mathrm{Im}(\omega) + i\mathrm{Re}(\omega)$. As can be seen, the new Hilbert transformed real and imaginary parts of the linear FRF will return the same circle line $\mathrm{FRF}(\omega) = \mathrm{Re}(\omega) + i\mathrm{Im}(\omega)$. Thus, the FRF of a linear system is invariant under the HT. But in the case of a nonlinearity the Hilbert transformed real and imaginary parts can have another form: $H\left[\mathrm{FRF}_{nl}(\omega)\right] = -\mathrm{Im}_{nl}(\omega) + i\mathrm{Re}_{nl}(\omega)$. So if the original FRF and its HT do not match, then the vibration system is most likely nonlinear. The level of distortion suffered in passing from the FRF to the HT can be given by the following complex function (Worden and Tomlinson, 2001) $\Delta\mathrm{FRF} = H\left[\mathrm{FRF}_{nl}(\omega)\right] - \mathrm{FRF}_{nl}(\omega) = \mathrm{Re}_{nl}(\omega) - \mathrm{Im}_{nl}(\omega) + i\left[\mathrm{Re}_{nl}(\omega) - \mathrm{Im}_{nl}(\omega)\right]$.

This way the HT allows nonlinearity to be detected on the basis of two measured vectors – the real and imaginary parts of the FRF. In practice the measured FRF is presented by a sampled data with digital frequency values, so numeric methods of the HT should be applied. Let us consider a forced vibration of a linear vibration system under a constant force amplitude excitation with a sweeping frequency. The measured synchronous input excitation and the output vibration allow us to estimate the transfer function – for example, by the MATLAB® function TFESTIMATE based on Welch's averaged periodogram method. The real part of the FRF plotted against its imaginary part, with a frequency as an implicit variable, is shown in Figure 8.3a. It is clear that the bene t of using a Nyquist plot comes from the circularity of the FRF in the complex plane. Once an accurate FRF is estimated, nding the HT becomes straight forward. For example, the HT can be applied to the digital data by the MATLAB function HILBERT, which produces almost the same circled FRF for the linear system (Figure 8.3a).

Now consider another case of a forced vibration of a nonlinear hardening cubic stiffness vibration system under the same force excitation with a sweeping frequency.

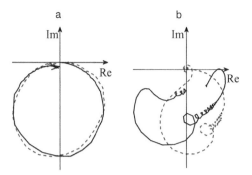

Figure 8.3   A Nyquist plot of the FRF: the linear vibration system (a), the nonlinear Duffing system (b); the measured data (−−), the Hilbert transformed data (—) (Feldman, ©2011 by Elsevier)

The resultant FRF is now distorted and the correspondence between $Re(\omega)$ and $Im(\omega)$ is no longer as good (Figure 8.3b). When the HT is applied to the digital FRF of a nonlinear system, the obtained Nyquist plot is distorted even more – by rotating clockwise and elongating into a more elliptical form (Figure 8.3b). Distortions due to other types of nonlinearities (e.g., softening cubic stiffness and Coulomb friction) are discussed in Worden and Tomlinson (2001). This example shows that the HT is a sensitive indicator of nonlinearity in the Nyquist plot resulting in a distorted version of the original FRF. A major problem in using the digital HT on FRF data occurs when a non-baseband (i.e., data that does not start at zero frequency) or a band-limited data is employed. The problem is usually overcome by adding correction terms to the HT evaluated from the measured data (Worden and Tomlinson, 2001).

   The HT analysis in the frequency domain is a nonlinearity diagnostic tool that is more sophisticated than other well-known techniques, such as the test for homogeneity when severe distortions are appearing in the FRFs as excitation level increases, or the test for isochrones when the frequency of motion becomes amplitude dependent. However, experience shows that the HT in the frequency domain allows us to detect, and sometimes to qualify, the type of nonlinearity (Bruns, Lindner, and Popp, 2003). For a quantitative identi cation of the system's nonlinear characteristics, we need to nd more advanced and more practical methods (Worden and Manson, 1998). Further developments of the advanced HT methods were focused on the time domain signal and system analysis. In this new approach, time becomes a central variable characterizing the input–output behavior of the vibration system. In later chapters the emphasis is not on the frequency domain; the discussion concentrates mainly on the methods in the HT time domain.

8.3   Typical nonlinear elasticity characteristics

A vibration motion occurs when a system has some sort of restoring (elastic, spring, stiffness) force which tends to move the mass to an initial equilibrium position. The mass moving back to the equilibrium position acquires a momentum that keeps it moving beyond that position, establishing a new restoring force in the opposite direction. This back and forth transfer of kinetic energy in the spring mass and potential energy causes the mass to oscillate. When an external force is removed – in a conservative (undamped) autonomous nonlinear system – the system will oscillate with the frequency of free vibration. This free vibration is the system's natural response; consequently, the free vibration frequency is referred to as the system's natural frequency. In nonlinear vibration systems the natural frequency can depend on an initial state, or a displacement level.

As a rst approximation we can assume that the system is linear and that the undamped natural frequency is approximately equal to the damped natural frequency since the damping is light. Because the IF of free vibration is the rst derivative of the phase angle, it was immediately suggested that an estimate should be made of the slope (the tangent) of the unwrapped phase line as an approximate natural frequency value. The slopes of an envelope and phase straight lines can be estimated using a linear least-squares t procedure, for example Braun (1986). Such a use of slopes of the envelope and phase function is a simple and an effective procedure (Yang, Kagoo, and Lei, 2000; Yang et al., 2003a; Salvino, 2000; Giorgetta, Gobbi, and Mastinu, 2007). However, it is acceptable for a linear vibration system only.

In reality, nonlinear stiffness is an important factor in causing the nonlinearity of vibration systems. This important type of nonlinearity arises when a restoring force is not proportional to the deformation in the system. There are several types of static load–displacement characteristics corresponding to some well-known nonlinear elasticity – such as backlash, precompressed, impact, and polynomial types. A vibrating system could also have a piecewise-linear restoring force that may be considered as approximations to continuous typical curves (Worden and Tomlinson, 2001). If the system to be tested has nonlinear symmetric elastic forces, the natural frequency will, in most cases, decisively depend on the amplitude of the vibrations. Most of the typical nonlinearities in a spring have a speci c form of skeleton curve (backbone) (Feldman and Braun, 1993). The following analysis of the topography of a skeleton curve is essential for the evaluation of the given properties in a vibrating system, for example, in reconstructing the characteristics of the elastic forces. Let us consider some typical cases of a nonlinear elastic force in a SDOF vibration system.

8.3.1   Large amplitude nonlinear behavior. Polynomial model

We know, best of all, the cases where nonlinearity occurs in large amplitude oscillations of elastic systems, for instance, a nonlinear spring element with a hard or soft restoring force (Figure 8.4a), or a nonlinear friction quadratic or cubic force. Whereas the amplitudes of vibration are large, the occurrence of these spring or damping nonlinearities cannot be ignored.

These typical nonlinear spring elements of a mechanical vibration system could be expressed as a power series $k(x) = (\alpha_1 + \alpha_3 x^2 + \alpha_5 x^4 + \ldots)x$. As will be shown in Section 8.6, an average natural frequency will form a corresponding backbone function $\omega^2(A) \approx \alpha_1 + \frac{3}{4}\alpha_3 A^2 + \frac{5}{8}\alpha_5 A^4 + \ldots$ almost repeating the same expression for the initial nonlinear restoring force $k(x)$ (Feldman, 1997). The nonlinear backbone skews toward a higher frequency in the case of a hardening spring (positive nonlinear

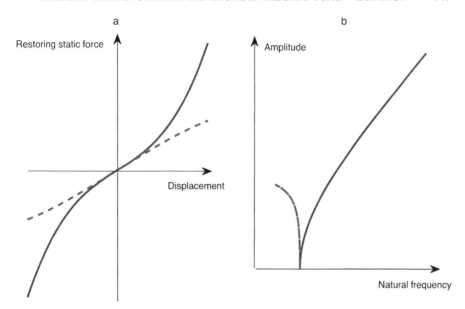

Figure 8.4 A polynomial nonlinear restoring force (a) and the vibration system backbones (b): a hardening spring (—), a softening spring (– –)

elastics members) and toward a lower frequency in the case of a softening spring (negative nonlinear elastics members) (Figure 8.4b).

8.3.2   Vibro-impact model

In the case when a nonlinear vibration system has absolute hard stoppers, the motion is possible only until the mass contacts the restriction (Figure 8.5) (Babitsky and Krupenin, 2001): $k(x) = \begin{cases} \omega_0^2 x, |x| < x_{max} \\ \pm\infty, \; x = \pm|x_{max}| \end{cases}$ Any further attempts to continue the motion over the rigid boundary will increase the elastic force and the natural frequency $\omega_0^2(A) = \begin{cases} \omega_0^2, A < x_{max} \\ \infty, A = x_{max} \end{cases}$. The corresponding ideal static force characteristics and the backbone of the vibro-impact (rigid boundary) system are shown in Figure 8.5.

8.3.3   Restoring force saturation (limiter)

For this idealization the static force is linear – up to the elastic force limit of the type of saturation (Figure 8.6a) $k(x) = \begin{cases} \omega_0^2 x, \quad |x| < x_{max} \\ \pm\omega_0^2 x_{max}, |x| \geq x_{max} \end{cases}$. For low excitation forces it leads to a linear behavior of the system when the natural frequency resides in the linear stiffness regime. After a certain point the saturation regime is reached, and the resonance frequency of the system decreases for rising displacements (Figure 8.6b) $\omega_0^2(A) = \begin{cases} \omega_0^2, \qquad A < x_{max} \\ x_{max}\omega_0^2/A, A \geq x_{max} \end{cases}$.

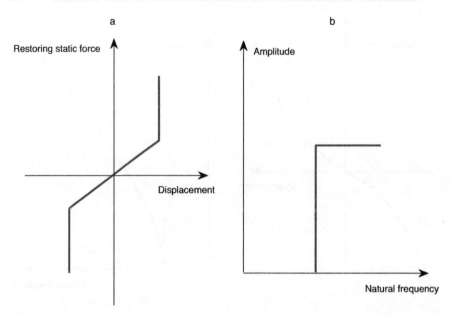

Figure 8.5   A vibro-impact nonlinear restoring force (a) and the vibration system backbone (b)

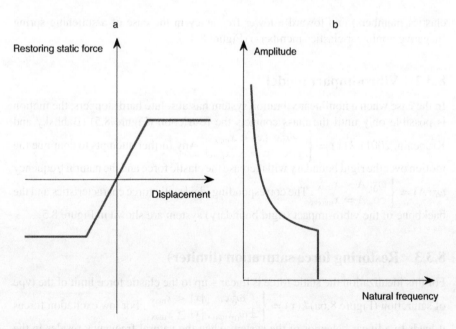

Figure 8.6   An elasticity saturation restoring force (a) and the vibration system backbone (b)

8.3.4   Small amplitude nonlinear behavior. Backlash spring

There are other cases when vibration systems show their speci c nonlinear behavior
in only a small amplitude range of vibrations. Such a system is, for example, a spring
backlash (clearance) (Worden and Tomlinson, 2001). The backlash vibration system
has a mass located in a gap Δ (clearance or dead zone) between two linearly elastic
springs. The static force characteristics have a typical form with a zero horizontal
zone $k(x) = \begin{cases} 0, & |x| \leq \Delta \\ \omega_0^2 (x - \Delta) \operatorname{sign}(x), & |x| > \Delta \end{cases}$ (Figure 8.7). The natural frequency of
the system with backlash depends on a gap magnitude, as well as on other parameters
(Weaver, Timoshenko, and Young, 1990): $\omega_0(A) = \dfrac{\omega_0}{1 + \frac{2}{\pi(A/\Delta - 1)}}$, $A > \Delta$. The backlash system backbone is a monotonically increasing curve that has a trivial vertical
line as an asymptote on the right (because of a constant natural frequency of the
corresponding linear system without a backlash), and cuts off a clearance value on
the amplitude axis on the left (Figure 8.7). The backlash system will mainly display
its nonlinear properties for small vibration amplitudes where a natural frequency
decreases extensively with each decrease of amplitude.

8.3.5   Preloaded (precompressed) spring

Another typical example of nonlinearity in a small amplitude range is a mechanical vibration with a preloaded (precompressed, pretensioned) restoring force. In
this model the spring in the equilibrium position is already precompressed by the

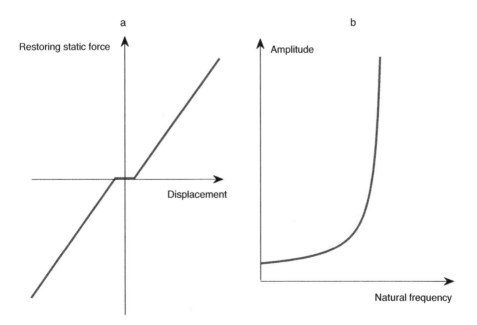

Figure 8.7   Restoring force with backlash (a) and the vibration system backbone (b)

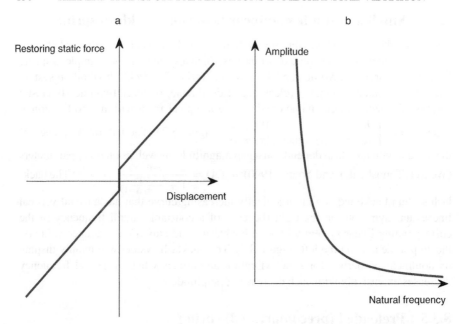

Figure 8.8 A precompressed restoring force (a) and the vibration system back-bone (b)

amount of a force F, as indicated by the static load–displacement diagram in Figure 8.8a: $k(x) = \begin{cases} \omega_0^2 x + F_0 \text{sgn}(x), & |x| > 0 \\ 0, & |x| = 0 \end{cases}$. The natural frequency in the presence of a precompresssed force decreases extremely for small vibration amplitudes:

$$\omega_0(A) = \frac{\pi \omega_0}{2 \arccos \frac{1}{1+A\omega_0^2/F_0}}. \tag{8.2}$$

The system backbone is arranged to the right of the asymptote $\omega_0^2(A) = \omega_0^2$, which corresponds to the linear system excluding the precompressed force. Actually, the natural frequency for large amplitudes is not dependent on a vibration amplitude. The precompressed deformation will extremely decrease the natural frequency only for small amplitudes.

8.3.6   Piecewise linear spring bilinear model

In some cases mechanical vibration systems have nonlinear or multivalued springs that results in piecewise linear restoring force characteristics. The simplest type of piecewise linear restoring force is a bilinear model in which the ﬁrst linear part at the origin has a slope $k_1 = \omega_{01}^2$ and the following linear part has a different slope $k_2 = \omega_{02}^2$: $k(x) = \begin{cases} \omega_{01}^2 x, 0 \le |x| < x_1 \\ \omega_{02}^2 x, \quad |x| \ge x_1 \end{cases}$. When a vibration amplitude A is less than x_1, the motion is simply harmonical and the corresponding segment of a skeleton

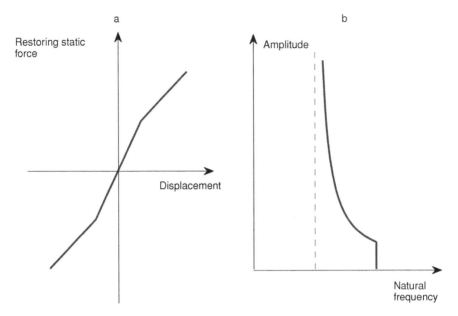

Figure 8.9   A bilinear restoring force (a) and the vibration system backbone (b)

curve is a trivial vertical line (Figure 8.9). But if the vibration amplitude is larger than x_1, the backbone will look like a typical curve for the backlash or saturation system.

The HT substitution of the primary solution establishes direct relationships between the parameters of the initial differential equations and the average instantaneous amplitude and natural frequency of the vibration response. The HT reduction allows a direct construction of an approximate solution to be de ned as a single quasiharmonic with a slow-varying amplitude and frequency (see Sections 9.2, 9.4).

8.3.7   Combination of different elastic elements

A real vibrating system could have several nonlinear, parallel-acting springs, whose equivalent spring force is the sum of the forces of the individual models, when all the springs have the same deformation. In general, the task to decompose the obtained backbone as a sum of typical curves has no unique solution. Nonetheless, when each spring model acts within its own amplitude zone, it is possible to represent the total backbone in the form of summed-up typical backbones.

8.4   Phase plane representation of elastic nonlinearities in vibration systems

The vibration motion of a SDOF nonlinear conservative system is described by the differential equation

$$m\ddot{x} + K(x) = 0 \tag{8.3}$$

where $m, \ddot{x}$, and $K(x)$ are, respectively, the mass, acceleration, and restoring force as a function of the displacement x. The restoring force can contain a linear stiffness and also any additional nonlinear restoring force component. The corresponding second-order differential equation of a conservative system for a unit mass will then take the general form

$$\ddot{x} + k(x) = 0 \qquad (8.4)$$

where the term $k(x) = K(x)/m$ represents the restoring force per unit mass as a function of the displacement x.

Traditionally, we introduce a new variable x, so that we can exclude time from the equation of motion although x and x are still time dependent, so $\ddot{x} = \frac{dx}{dt} = \frac{dx}{dx}x$. Using the new variable x is a traditional way of studying the motion of an oscillator by representing this motion in the x, x plane (Figure 8.11), where x and x are orthogonal Cartesian coordinates (Andronov, Vitt, and Khaikin, 1966). In the new coordinates, the last equation takes the form: $\frac{dx}{dx} = \frac{k(x)}{dx}$. After variable separation and integration, the phase plane takes the form $\frac{1}{2}x^2 - \int k(x)dx = C$, where C is a constant. After the rst integration the last equation yields an expression for the velocity x of the unit vibrating mass in any position; after the second integration it yields the time of a full oscillation cycle (Weaver, Timoshenko, and Young, 1990). Hence, the time of the full cycle becomes

$$T = 4 \int\limits_{0}^{x_{max}} \frac{dx}{\sqrt{2 \int_{x}^{x_{max}} k(x)dx}}, \qquad (8.5)$$

where x_{max} is the maximum value of displacement, therefore the velocity corresponding to x_{max} in an extreme position is zero. The time of the full cycle taken for an oscillation to occur completely is referred to as the oscillatory period. So for a linear restoring force that will result in a constant period value equal to 2π, the integral curve of the linear harmonic oscillator will be replaced by a circle in the phase plane, representing a rotating point with a constant angular velocity.

In the classic theory of vibration, the full period of oscillation is determined between the two instant local values of displacement (for example, at zero or extreme position). This precise full period value for a nonlinear restoring force can depend on the maximum amplitude (the total stored energy) and is called the nonisochronism of a nonlinear oscillation. To compute the value of the period, we need to know where the object was at the beginning (x_0) of the motion, and where it is at the end (x_{max}) of it. Obviously, we do not need to know what was happening in the middle of the vibration motion.

However, we can consider an actual period of oscillation as a continuous function of time that can vary in time during a small part of the oscillation (Braun and Feldman, 1997). It will represent the expression for the period T not only as the full cycle value, but also as a continuous function of motion x (8.5). In consequence, we will get a function of the *instantaneous period* with possible fast intramodulations during any short part of the period $T(t)$. By dividing the phase plane into small sectors and estimating the integral (8.5) separately for each small interval, we will get the instantaneous period of oscillation $T(t)$ as a function of the instant position. So, for a nonlinear system, the instantaneous period will become a varying intrafunction of the phase angle. As a result of the full cycle integration of the instantaneous period, we will get back the classic precise full period value.

In the classic $x, \dot{x}$ plane (Figure 8.11), where x and $\dot{x}$ are orthogonal Cartesian coordinates, the inverse function of an instantaneous period can be called the *instantaneous natural angular frequency* $\omega(t) = 2\pi \, [(T(t)]^{-1}$ (Braun and Feldman, 1997). The instantaneous natural frequency for a nonlinear system can be a fast-varying function of time and, respectively, the instantaneous radius vector in the phase plane $r(t) = \sqrt{x^2(t) + \dot{x}^2(t)}$ will also vary during a single cycle from its minimum up to its maximum value. This means that the instantaneous natural frequency of nonlinear systems is not just a number of cycles, but is a continuously varying function of time. In a similar way, fast intrawave modulations during any short part of the natural period have been discussed in Huang *et al.* (1998).

To illustrate these intrawave variations of the instantaneous natural frequency, let us consider an example of the Duf ng equation with an initial nonzero displacement and without damping $\ddot{x} + (1 + \varepsilon x^2)x = 0$, where $\varepsilon = 5$, $x_{\max} = 1.0$. The obtained free vibration solution of the Duf ng equation, and the calculated instantaneous period, are shown in Figure 8.10. As can be seen in Figure 8.10b, the instantaneous

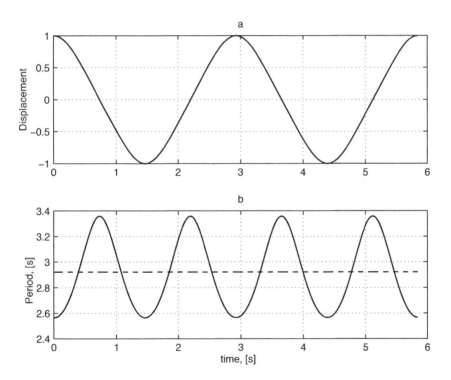

Figure 8.10 The solution (a) and instantaneous oscillation period (b, —) of the Duffing equation: the average period (−−) (Feldman, ©2011 by Elsevier)

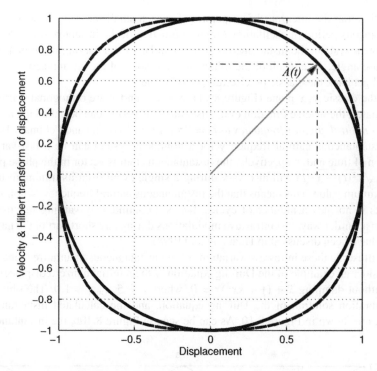

Figure 8.11   The Duffing equation ($\varepsilon = 5$) phase plane (– –) and the analytic signal (—) (Feldman, ©2011 by Elsevier)

period is a varying function of time and oscillates two times faster than the nonlinear solutions. It oscillates with a deviation from the average value of more than 10% (Braun and Feldman, 1997).

The instantaneous radius vector in this case also becomes a varying function $r(t) = \left(-0.5\varepsilon x^4 + x_{max}^2 + 0.5\varepsilon x_{max}^4\right)^{1/2}$ and the portrait mapped in the phase plane differs from the ideal circle (Figure 8.11). It is clear that the traditional direct integration for the total period (8.5) produces only a period value T, which corresponds to the total cycle of the motion. In a general case of nonlinear systems, all instantaneous parameters oscillate with fast frequencies during each cycle of vibration. The appearance of these fast oscillations, from the point of view of time, is the effect of the existence of an in nite number of high ultra-harmonics in addition to the fundamental (primary) quasiharmonic solution.

8.5   Complex plane representation

Let us consider another important technique for a presentation of a vibration solution on the basis of the rotating vectors. According to the analytic signal theory, a real vibration process $x(t)$, measured by, say, a transducer, is only one of many possible projections (the real part) of some analytic signal $X(t)$. The second projection of the same signal (the imaginary part) $\tilde{x}(t)$ will then be conjugated according to the HT (Hahn, 1996a). The notion of the analytic signal, introduced by D. Gabor, is viewed as a complex-valued time function provided by a joint analysis of the initial signal and that conjugated according to the HT. Each point in the complex plane is characterized by a radius vector $A(t)$ and a polar angle $\varphi(t)$ (Figure 8.11). As a matter of fact, the analytic signal presentation is nothing other than a development of the complex amplitude method, when the signal magnitude (envelope) and phase parameters vary in time and can be determined by the HT as single-valued instantaneous functions (Adamopoulos, Fong, and Hammond, 1988; Vakman, 1998). Therefore, the analytic signal method is an expedient that is effective enough to solve the general problems of vibration theory, such as the analysis of free and forced nonstationary vibrations (transient processes) and nonlinear vibrations (Vainshtein and Vakman, 1983). The analytic signal representation allows us to consider any vibration at any instant of time t – as a quasiharmonic oscillation – both the amplitude and the angle modulated functions:

$$x(t) = A(t)\cos\varphi(t) = A(t)\cos\int\omega(t)dt, \qquad (8.6)$$

where $\omega(t) = \varphi(t)$ is the IF of the vibration.

To illustrate the analytic signal representation, once again consider an example of the same Duffing equation with a constant initial amplitude and without damping $\ddot{x} + (1 + \varepsilon x^2)x = 0$, where $\varepsilon = 5$, $x_{\max} = 1.0$ (Figure 8.12).

As can be seen, the envelope and the IF are fast varying functions and they oscillate two times faster than the nonlinear solution $\omega(t) = \omega_0 t + 2\omega_0\varepsilon_0\cos 2\omega_0 t$. The doubled modulation frequency according to (4.2) produces at least two spectral components with a main frequency and tripled frequency values $x(t) = A\cos(\omega_0 t + \varepsilon_0\sin 2\omega_0 t) \approx A\cos\omega_0 t + \varepsilon\cos 3\omega_0 t$. The time modulation frequency then gives rise to additional multiple spectral components of the solution. The *intrawave* frequency *modulation* that indicates a nonlinear vibration behavior can be observed in the time domain as modulation. The same phenomenon in the frequency domain can be denoted as energy contributions in the additional spectrum peaks. The *intrawave* frequency *modulation* is therefore depicted in the HT analysis as well as in the Fourier or Wavelet analysis.

The modulation of the instantaneous parameters takes place for both mathematical interpretations: for the classic phase plane and for the analytic signal representation. Thus, the appearance of fast oscillations in the solution simply reflects the nonlinear nature of vibration systems; they are not associated with the interpretation of the HT signal. This permits the nonlinear solution to be represented in an asymptotic expansion form (Section 5.2) $x(t) \approx \sum_1^k A_l(t)\cos\varphi_l(t)$ with a set of multiple frequency components, where every component of the solution has slow-varying parameters. In this way the solution of nonlinear systems will include high-frequency superharmonic components in addition to the primary harmonic.

The obtained instantaneous amplitude and the instantaneous natural frequency – as well as their relationship dependency (the envelope vs. IF function) – can have

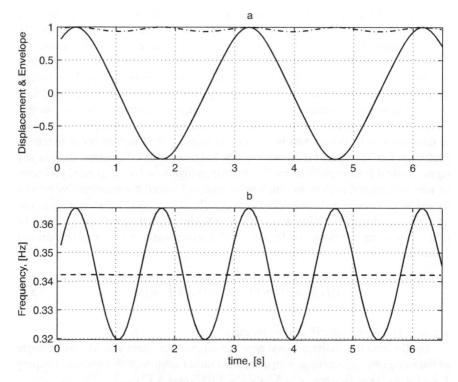

Figure 8.12   The solution of the Duffing equation (a, —), the envelope (a, ···), the IF (b, —), and the average IF (b, – –) (Feldman, ©2011 by Elsevier)

an unusually fast oscillation (intrawave modulation) regardless of the traditional phase plane (x, x) or complex plane $(x, \tilde{x})$ representation of a nonlinear system. This naturally occurring fast modulation requires a more sophisticated analysis, and will be further investigated in detail in Chapter 11.

8.6   Approximate primary solution of a conservative nonlinear system

In general, the fast-varying solution of the second-order conservative system (8.4) with a nonlinear restoring force has the form $x(t) = A(t) \cos \int \omega(t) dt$. The nonlinear restoring force $k(x)$ as a function of time can be transformed into a multiplication form of a new varying nonlinear natural frequency $\omega^2(t)$ and system solution $k(x) = \omega^2(t)x(t)$.

To apply the multiplication property of the HT (2.3), let us assume that the new varying nonlinear natural frequency squared could be grouped into two different parts: $\omega^2(t) = \overline{\omega_0^2}(t) + \omega_1^2(t)$. As the rst part $\overline{\omega_0^2}(t)$ is much slower and the second component $\omega_1^2(t)$ is faster than the system solution, the equation of motion will get a new form: $\ddot{x} + \left[\overline{\omega_0^2(t)} + \omega_1^2(t)\right]x(t) = 0$. Now, according to the multiplication property of the HT for overlapping functions (2.3), we apply the HT for both sides of the last equation $\ddot{\tilde{x}} + \overline{\omega_0^2(t)}\tilde{x}(t) + \widetilde{\omega_1^2}(t)x(t) = 0$, where the tilde sign indicates the conjugate variable of the HT. Multiplying each side of the obtained HT equation by i and adding it to the corresponding sides of the initial equation, we obtain a new differential equation of motion in an analytic signal form $\ddot{X} + \overline{\omega_0^2}X + \left(\omega_1^2 + i\widetilde{\omega_1^2}\right)x = 0$, where X is a complex solution in an analytic signal form: $X = x + i\tilde{x}$. This complex equation can be transformed into a more traditional and accepted form (Feldman, 1997):

$$\ddot{X} + i\delta\dot{X} + \omega^2 X = 0, \tag{8.7}$$

where $\omega^2 = \overline{\omega_0^2} + \dfrac{\omega_1^2 x^2 + \widetilde{\omega_1^2}x\tilde{x}}{A^2};\ \ \delta = \dfrac{\widetilde{\omega_1^2}x^2 - \omega_1^2 x\tilde{x}}{A^2}$. Here ω^2 is a varying instantaneous natural frequency and δ is a *fast-varying instantaneous fictitious friction* parameter. It should be pointed out that this equation is not a real equation of nonlinear motion; however, it produces the same varying solution of the system. Equation (8.7), based on the HT, de nes the instantaneous natural frequency as a constantly time-varying function, but not as the number of occurrences during one period of time. It is just an arti cial ctitious equation that produces the same nonlinear primary vibration solution. The varying natural frequency ω^2 (8.7) obtained consists of a slow $\overline{\omega_0^2}$ and a fast intrawave component. The ctitious friction parameter δ, on the other hand, consists only of a fast-varying component. This means that the slow part of the natural frequency forms an average period of vibration, but the fast-varying ctitious friction force (induced by the zero average value) has no effect on the real average friction force. The results explain the experimental fact that – according to the HT analysis of nonlinear systems – all the instantaneous characteristics, such as the IF and the amplitude of the solution, take an unusually fast oscillation (modulation) form (Davies and Hammond, 1987).

Expression (8.7) allows us to nd a general relation between the initial nonlinear static force characteristics $k(x)$ (8.4) and the varying instantaneous natural frequency ω^2. The nonlinear restoring force can be represented by an expansion with a power series form (Nayfeh, 1979)

$$k(x) = (\alpha_1 + \alpha_3 x^2 + \alpha_5 x^4 + \ldots)x = \left(\sum_{n=1}^{N} \alpha_{2n-1}x^{2n-2}\right)x. \tag{8.8}$$

After substituting the expanded restoring force characteristics with the primary solution $x(t) = A(t)\cos\varphi(t)$ into the varying natural frequency (8.7), one can derive the particular type of nonlinear natural frequency as a function of time $\omega_0^2(t) = \mathbf{F}[A(t), \varphi(t)]$. By integrating the varying instantaneous natural frequency function of time ω^2 from (8.7) on the interval $[0\ T]$, where T is the full period of the primary solution, we will get an average natural frequency function:

$$\langle\omega^2(A)\rangle = T^{-1}\int_0^T \omega^2(t)dt = \alpha_1 + \frac{3}{4}\alpha_3 A^2 + \frac{5}{8}\alpha_5 A^4 + \ldots \tag{8.9}$$

The obtained average nonlinear natural frequency is strictly dependent on the primary solution amplitude. If the amplitude is a slow-varying function of time, then the average natural frequency will also become a slow-varying function of time.

Comparing two formulas (8.9) and (8.8) we will reach the important general conclusion that an average natural frequency simply repeats the structure of an initial nonlinear elastic characteristic, correct to the corresponding polynomial numeric coef cients: $\langle \omega^2(A) \rangle \Leftrightarrow k(x)$. In essence, the HT approach is just an alternative to the well-known linearization methods (like a harmonic balance linearization). The estimated average natural frequency function includes all the information about the initial system and could be used to identify a nonlinear system. A variation of an average natural frequency with an amplitude can be represented in the manner of a nonlinear *backbone (skeleton) curve*. This curve depicts the natural frequency as a function of the free vibration envelope, so it constitutes an inherent feature of nonlinear systems, showing that the frequency of motion is amplitude dependent.

For example, a nonlinear system described by the Duf ng equation with a linear and positive hard cubic spring $k(x) = \alpha_1 \left(1 + \alpha_3 x^2\right) x$ will produce a primary solution with an average natural frequency equal to $\omega^2(A) = \alpha_1 \left(1 + 3\alpha_3 A^2/4\right)$. This approximation of the average natural frequency (Figure 8.13, dotted line) maps an approximate backbone very close to the precise backbone (Figure 8.13, bold line).

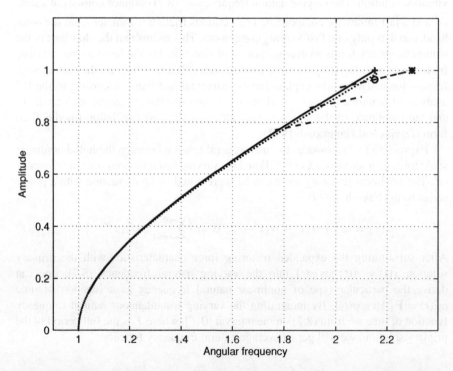

Figure 8.13 The backbone of the Duffing system ($\alpha_3 = \varepsilon = 5$): a precise period solution (—), the average natural frequency (. . .) (Feldman, ©2011 by Elsevier)

In this case, the IF mean value obtained through the HT is very close to the theoretical value of the free vibration frequency. This fact indicates that the HT realizes a good representation of nonlinear vibrations.

At rst glance, a stable conservative nonlinear vibration system with a nonzero initial displacement (velocity) traditionally should have a constant amplitude and a natural frequency. This means that oscillation of such a vibration system without damping should map on the frequency–amplitude (backbone) plot just a single point corresponding to the amplitude and frequency as though they were constants. However, due to the HT representation, which is able to show real variations of the instantaneous vibration characteristics – the amplitude (envelope) and the natural frequency – the vibration motion will map a closed drawn off gure compressed to a short tilted line. An example of such short lengths for three different initial amplitudes of the Duf ng equation without damping is shown in Figure 8.13 by the dashed lines. During every full vibration period a mapping point moves around the average values along the short length line from one end to the other, and back again.

8.7   Hilbert transform and hysteretic damping

A majority of real materials show an energy loss per cycle with a less pronounced dependency on frequency. In fact, many materials indicate force – deformation relations that are independent of the deformation rate amplitude — so-called hysteretic relations. Those known mechanical element models show frequency-independent storage and loss moduli, which implies that the Fourier transforms of both the element force $F(\omega)$ and the element deformation $\Delta(\omega)$ satisfy $F(i\omega) = K[1 + i\eta \, \text{sgn}(\omega)] \, \Delta(i\omega)$, where K is stiffness and η is a loss factor (a damping ratio of the loss and storage moduli of an element). In such a case, the differential equation that describes a free movement of a SDOF system becomes $m\ddot{x} + k(1 + i\eta)x = 0$, where η is the hysteretic damping ratio.

In this model, the dissipated energy in a harmonic deformation cycle of a constant amplitude is independent of the frequency of deformation. Only the HT gives a correct time domain expression for the concept of a linear hysteretic (Inaudi and Kelly, 1995). This transform can be used to replace complex-valued coef cients in differential equations modeling mechanical elements.

The HT also forms relations between the real and imaginary parts of the dynamic stiffness matrix of a causal hysteretic damping (Makris, 1999). Based on the analyticity of the transfer function, it was shown that a causal hysteretic damping model is a limited case of a linear viscoelastic model with an almost frequency-independent dissipation.

8.8   Nonlinear damping characteristics in a SDOF vibration system

Damping is a dissipation of energy from a vibration system. A damped vibration system without a permanent excitation responds by vibrating freely until it dissipates all of the potential and kinetic energy it received and stored. This dissipation, or damping, describes the physical phenomenon of converting the energy of motion into another form, mainly into heat and sound. In mechanical vibration this is accomplished mainly through a frictional resistance (Fidlin, 2005).

The damping force is assumed to be relative to the instantaneous velocity, but it is always directed against the velocity. Nonlinear frictional force could be described using different expressions; for example, an equation of motion for a SDOF model with a nonlinear viscous damping is $m\ddot{x} + C(x) + K(x) = 0$, where $m, C(x)$, and $K(x)$ are respectively the mass, the nonlinear frictional force as a function of velocity x, and the nonlinear restoring force as a function of displacement x. The corresponding second-order differential equation of the damped system for unit mass will then take the following form

$$\ddot{x} + 2h(x)x + \omega_0^2 x = 0 \tag{8.10}$$

where the term $h(x)x = C(x)/2m$ represents the damping force per unit mass as a function of velocity and the term $\omega_0^2 = k/m$ represents the undamped natural frequency squared in the case of the linear restoring force $K(x) = kx$. In control theory it is sometimes desirable to use a dimensionless quantity term ζ called the damping ratio, $\zeta = h/\omega_0 = c/2\sqrt{km}$, where h is a damping coef cient (factor) for a linear viscous damping showing the rate of decay $h(x) = hx$; and k is the linear stiffness from $K(x) = kx$.

Equation (8.10) can be solved by assuming a transient (unforced) quasiharmonic underdamped solution x such that $x(t) = A_0 e^{-h(t)t} \cos \omega t$, where $h(t)$ is an *instantaneous damping coefficient* and $\omega = 2\pi/T$ is an angular damped natural frequency, corresponding to the time spacing T between two close damped natural periods (Inman, 1994). The damped natural frequency of the solution is less than the undamped natural frequency ($\omega < \omega_0$).

Several consecutive peaks of a free response $x_{max}(t)$ during the time $t = Tn$ characterize an average relative damping in the system

$$x_{max}(Tn) = A_0 e^{-\langle h \rangle Tn}. \tag{8.11}$$

Classically, this average relative damping is measured by a *logarithmic decrement* as a natural logarithm of the peaks ratio

$$\delta = \frac{1}{n} \ln \frac{x_{i+n}}{x_i} = \langle h \rangle T = \frac{2\pi \langle h \rangle}{\omega}, \tag{8.12}$$

where δ is the logarithmic decrement, $\langle h \rangle$ is the average damping coef cient, T is a period, and ω is the angular frequency of the damped solution. The logarithmic decrement is a dimensionless term that is in close relation with the dimensionless damping ratio: $\zeta = \delta\omega/2\pi\omega_0 \approx \delta/2\pi$; $\delta \approx 2\pi\zeta$.

The plot of dependency between the amplitude and the damping coef cient $A(h)$ forms a *damping curve*. The particular shape of the damping curve differs for various damping mechanisms. For instance, in the case of a linear viscous friction $h(x) = h$, when the damping coef cient h is a constant and does not vary during free decay, the damping curve will be a vertical straight line. The envelope of free decay vibration is determined by the monotonic exponent decay rate. So the envelope is an exponential function of time $A(t) = A_0 e^{-ht}$. The envelope can be shown in a logarithmic amplitude scale where the exponential envelope curve $A(t)/A_0$, determined by the decay rate h, can be seen as a linear decay $\ln[A(t)/A_0] = -ht$ (Herlufsen, 1984; Hammond and Braun, 1986; Agneni and Balis-Crema, 1989).

The slope of an envelope straight line can be estimated using a linear least-squares t procedure, suggested for example in Braun (1986). We should assume that the system is linear and that the undamped natural frequency is approximately equal to the damped natural frequency since the damping is light. The IF is the rst derivative of the phase angle, so it was also suggested to estimate a slope (the tangent) of the unwrapped phase line as an approximate natural frequency value. Such a use of slopes of the envelope and phase function is a simple and but effective procedure (Yang, Kagoo, and Lei, 2000; Yang et al.,, 2003a; Salvino, 2000; Giorgetta, Gobbi, and Mastinu, 2007). However, it is only acceptable for a linear vibration system; the procedure of the logarithmic-decrement slope deals only with the integrated damping estimation, and not with the instantaneous damping parameters, as will be shown in Section 9.2. In Section 8.9 and Section 9.2, which is devoted to the identi cation of a nonlinear system, we will show how a nonlinear damping force shapes the envelope function of the free vibration solution.

8.9   Typical nonlinear damping in a vibration system

When dimensionless damping parameters, such as a logarithmic decrement or a damping ratio, depend on the driving frequency, the damping mechanism is frequency (velocity, rate) dependent. When they are not sensitive to the forced frequency variations, the damping mechanism is frequency independent. Thus, we have two different types of energy loss in a vibration motion: one frequency dependent and the other frequency independent. In the majority of real mechanical constructions, damping typically does not depend on frequency. This can be observed in MDOF vibration systems, when the damping ratio is almost the same for all natural modes with different frequencies.

Typically, vibration systems show the evidence of damping nonlinearities in either a small or a large range of amplitudes, depending on the type of nonlinear damping force characteristics. An example of a small amplitude nonlinear behavior is a system with Coulomb (dry) friction, whose plot of logarithmic decrement vs. vibration amplitude is a monotonic hyperbola (Feldman and Braun, 1993). In some other cases, such as nonlinear turbulent friction, nonlinear damping behavior appears over a large range of amplitudes. But, in practice, in all cases the real damping force level is much less than the elastic force. Therefore a nonlinear damping has almost no effect on natural frequencies and corresponding system backbones.

The instantaneous damping characteristics of a vibration system are determined by the form of a symmetric frictional force (dissipative function). When a vibration system has well-known typical nonlinear damping characteristics, the corresponding instantaneous damping parameters are also typical functions of the amplitude and/or the forced frequency. These dependencies allow us to estimate the type and value of different kinds of damping: dry friction, structural, viscous, and so on. Utilizing the analytic signal representations we can estimate the characteristics of not only the elastic force but also of the damping force. Any mathematical representation of physical damping mechanisms in the equations of motion of a vibration system is only an idealization and approximation of the true physical situation.

8.10   Velocity-dependent nonlinear damping

In speci c cases when a damping force depends on a frequency variation, as in a viscous nonlinear damping force model, the energy dissipation becomes a function of velocity. Since the vibration velocity is a multiplication of the amplitude and frequency, the damping force will directly depend on the vibration frequency. Historically, a nonlinear friction force was represented by an expansion having a power series form of the velocity x and signum function $\text{sgn}(x) = x/|x|$. The damping force in this case is

$$F_{\text{damping}} = 2h(x)x = 2h|x|^n x/|x| = 2hx|x|^{n-1}, \qquad (8.13)$$

where h is a damping proportionality constant, and n is a nonlinear index. For example, a velocity squared or turbulent damping $(n = 2)$ is present when a mass vibrates in a   uid or in the air with a rapid motion. A turbulent damping force has a form: $F_{\text{damping}} = 2hx|x|$. A viscous linear damping $(n = 1)$ is the simplest type of damping force $F_{\text{damping}} = 2hx$. Coulomb damping $(n = 0)$, for example, is used to represent a dry friction in sliding surfaces $F_{\text{damping}} = 2h\,\text{sgn}(x)$.

A nonlinear damping force $h(x)$, as a function of velocity in the equation of motion (8.10), can be transformed into a function of time in the form of the multiplication of the instantaneous damping coef cient $h(t)$ and the velocity: $h(x) = h(t)x(t)$.

Assume that the primary solution in a quasiharmonic form $x(t) = A(t)\cos\varphi(t)$. This allows us to separate the friction force function in time into two different parts: $h(t) = \bar{h}_0(t) + h_1(t)$. The   rst part $\bar{h}_0(t)$ is much slower, and the second component $h_1(t)$ is faster than the system solution. The separation will allow the use of the HT of the force function and a complex form of the equation of motion. As a result we can precisely identify the existing damping mechanism by estimating the damping curve and damping force characteristics. Analytically the dynamics of a model mechanical system with fast oscillations of the damping coef cient has been analyzed by Fidlin (2005). He performed an asymptotic analysis of the equation of motion of this system and concluded that these oscillations also produce variations in its effective stiffness.

It is well known that the particular damping curve shape differs for various damping mechanisms. For instance, in the case of a linear viscous friction $(n = 1)$, the damping coef cient is a constant $\langle h \rangle = h_0$ and does not vary during a free decay.

Let us analyze the damping force separately for typical classical nonlinear damping models.

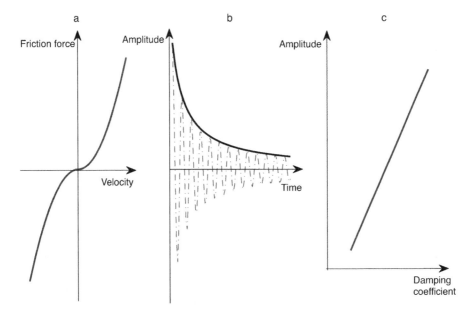

Figure 8.14   The turbulent quadratic damping model: the frictional force character-
istics (a), the envelope of a free decay (b), and the damping curve (c)

8.10.1   Velocity squared (quadratic, turbulent) damping

In the case of a quadratic turbulent damping, the index $n = 2$ and the damping
force has a parabolic shape for each velocity sign (Figure 8.14a). We can then
write the force function $F_{damping}(t) = -2h|A\omega \sin \varphi(t)|A\omega \sin \varphi(t) = -2h[2/\pi +$
$A\omega \sin \varphi(t) - 2/\pi]A\omega \sin \varphi(t)$, shown in Figure 8.15 by a dash-dot line. The same
gure shows the HT projection of the damping force (dashed line) and the damping

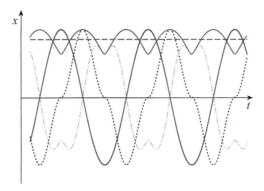

Figure 8.15   The turbulent damping force in time: the quasiharmonic solution (—),
the damping force (. . .), the HT of the damping force (-··-), the damping force envelope
(—), and the force envelope mean value (– –)

force envelope function (bold line). We will point out that the damping force envelope is a fast oscillation function. Such a fast oscillation of the damping force initiates fast intra-modulations of the solution during every period of oscillation.

By integrating the envelope of the varying damping force function over an interval $[0\ T]$, where $T = 2\pi/\omega$ is the full period of the primary solution, we will get an energy average value of the damping force $\langle F_{\text{damping}}(t)\rangle = h\left(8A\omega/3\pi\right)x$ (Chopra, 2007). The last expression indicates that, in the case of the turbulent damping, the average damping coefﬁcient is directly proportional to the envelope of the solution (Figure 8.14c):

$$\langle 2h\rangle = h_0 A\omega, \quad \text{where } h_0 = h\left(8/3\pi\right). \tag{8.14}$$

In essence, the HT approach is one more alternative to well-known equivalent damping energy methods.

It is interesting that the time-domain representation of motion equation solutions allows us to exploit relations between the damping coefﬁcient, the envelope, and the ﬁrst derivative of the envelope as a function of time $2h(A) = -\dot{A}/A$ – as will be shown in Section 9.2. A separation of the variables A and $\dot{A}$ – for the turbulent damping – gives a simple differential equation that has the form $\frac{dA}{A} = -h_0 A\omega d(t)$. The obtained equation can be integrated $\int_{A_0}^{A}\frac{dA}{A^2} = -h_0\omega\int_0^t d(t)$ with the following solution $A(t) = A_0/(1 + A_0 h_0\omega t)$, so the free decay envelope of the turbulent damping decreases like a hyperbola. The free decay amplitude decreases even faster than the trivial exponent of linear damping (Figure 8.14b). The average damping coefﬁcient is proportional to the vibration amplitude (8.14); its value increases with increasing amplitude $\langle h\rangle \sim A\omega$. The turbulent damping shows its nonlinear behavior mostly for large amplitude values.

8.10.2   Dry friction

Coulomb (dry) friction is a kind of nonlinear damping when the force resisting the motion is assumed to be proportional to the normal force between the sliding surfaces and is independent of velocity, except for the sign (Figure 8.16a). Thus, the damping force is $F_{\text{damping}} = 2h\,\text{sgn}\,(\dot{x})$. Assuming a primary solution in quasiharmonic form $x(t) = A(t)\cos\varphi(t)$ we can show the damping force as a square wave (Figure 8.17, dash-dot line).

Notice that the HT of the signum function is of the form: $2hH[\text{sgn}(\sin\varphi)] = 2h2/\pi \ln|\tan(\varphi/2)|$ (King, 2009). Figure 8.17 also demonstrates the HT projection of the damping force (dashed line) and the damping force envelope function (bold line). Again, we will point out that the damping force envelope is a fast-oscillation function. Such a fast oscillation of the damping force initiates fast intra-modulations of the solution during every period of oscillation (see Figure 8.17).

By integrating a varying envelope of the damping force over an interval $[0\ T]$, where $T = 2\pi/\omega$ is a full period of the primary solution, we will get an energy average value of the damping force $\langle F_{\text{damping}}(t)\rangle = h\left(4/\pi A\omega\right)x$. The last expression indicates that the average damping coefﬁcient in the case of a dry friction is inversely proportional to the envelope of the solution: $\langle 2h\rangle = h_0/A\omega$, where $h_0 = h\left(4/\pi\right)$. So

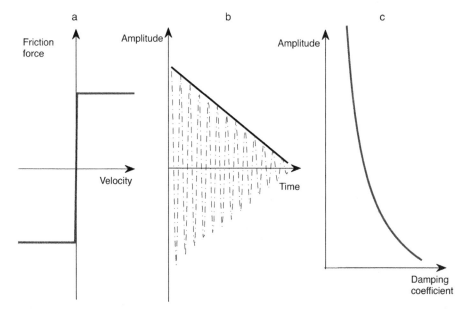

Figure 8.16   A dry friction model: the friction force (a), the envelope of a free decay (b), and the damping curve (c)

Coulomb, or dry friction in particular, has a plot of the damping curve as a monotonic decreasing hyperbola (Figure 8.16c).

In Section 9.2 we will show that during a free vibration regime both the envelope and the rst derivative of the envelope, as functions of time, are related to the instantaneous damping coef cient: $2h(A) = -A/A$. Separating the variables, a simple differential equation can be written in a rearranged form: $\frac{dA}{A} = -(h_0/A\omega)d(t)$. The

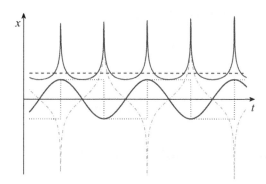

Figure 8.17   The dry friction force function in time: the quasiharmonic solution (—), the square damping force (. . .), the HT of the damping force (---), the damping force envelope (—), and the force envelope mean value (--)

obtained equation can be integrated $\int_{A_0}^{A} dA = -\frac{h_0}{\omega} \int_0^t d(t)$ with the following solution $A(t) = A_0 - \frac{h_0}{\omega} t$. The result shows that the free decay envelope of the dry friction decreases linearly in the form of a straight line (Figure 8.16b).

A vibration system with a dry friction force (Figure 8.16a) corresponds to the index $n = 0$ (8.13) when the amplitude of a free decay decreases linearly (Figure 8.16b), and the damping curve has a typical monotonic hyperbolic form (Figure 8.16c). The presence of dry friction becomes visible mostly for small vibration amplitudes when both the damping coefficient and the logarithmic decrement extremly increase.

8.11   Velocity-independent damping

Most materials show an energy loss per cycle with a very small dependency on frequency. In fact, many materials indicate force–deformation relations that are independent of the deformation rate amplitude – so-called hysteretic, structural, or internal damping relations. The known mechanical element models show frequency-independent storage and loss modulus. This implies that the damping force per unit mass in (8.10) is a function of the HT conjugate projection of the displacement $m\ddot{x} + i\eta(x)x + kx = 0$. In this case, the dissipated energy in a harmonic deformation cycle of a constant amplitude is independent of the frequency of deformation. As already mentioned, only the HT gives a correct time-domain expression for the concept of a linear hysteretic (Inaudi and Kelly, 1995). This transform can be used to replace complex-valued coefficients in differential equations modeling mechanical elements.

The damping parameters would depend differently on the amplitude according to the action of the nonlinear structural damping. For example, the case of $\eta(x) = \eta_0$ corresponds to the linear structural friction when the damping coefficient is a constant, and does not vary during free decay. Such linear structural friction produces a trivial exponential free decay plot. In nonlinear cases the amplitude of a free decay will be a more complicated function of time. The case of a dry friction $n = 0$ (8.13), when the damping curve has a hyperbolic form (Figure 8.16c), also corresponds to a typical case of structural damping.

Usually the frequency of the main solution of nonlinear systems under a free vibration regime varies only moderately. Such a small variation in frequency does not allow us to analyze the dependency between damping and frequency. Traditionally, to discover whether the damping is of a frequency-dependent or frequency-independent nature, it must be tested under a forced vibration regime.

8.12   Combination of different damping elements

Naturally, damping is a complex phenomenon and more than one damping type may exist in the same real structure. Considering the total frictional force as a sum of typical frictional elements, one can write an expression for the total damping coefficient as a sum of the typical dependencies. However, the decomposition of this total damping coef cient has no simple solution. It is only when each simple damping mechanism operates in an appropriate different range of amplitude and/or forced frequency that we can obtain a unique interpretation of the general damping dependence.

The instantaneous damping coef cient can be determined in the form of a symmetric dissipative function. In the particular case of a linear system, the instantaneous natural frequency and instantaneous damping coef cient, or decrement, do not vary in time. In the general case of nonlinear systems, the instantaneous damping coef cient and the natural frequency become amplitude and frequency functions.

If nonlinear dissipative forces are operating in a vibration system, the instantaneous damping coef cient may depend on the instantaneous amplitude. Experimental studies of the vibration of engineering structures indicate that the nature of dissipative forces is such that the frequency has almost no in uence on the value of a logarithmic decrement, and that a model of frequency-independent friction should be used to describe the vibrations. Any mathematical damping model obtained from an experiment will be only an idealization; it does not give a detailed explanation of the underlying physics.

The real damping in a mechanical construction is a summation of several components. Various nonlinear damping mechanisms of structural components can be realized arti cially in real physical models (Mai *et al.*, 2008). This realization is caused by the various types of nonlinearities that can be generated by varying the connection details of timber joints.

8.13   Conclusions

In many practical cases, measured mechanical vibration is not small and nor is the system linear. In these cases a nonlinear analysis becomes essential to understand the physics of the system or of the signal. The chapter is focused on the effects connected with the nonlinearities encountered in mechanical engineering. A Hilbert transform analysis clearly displays the main feature of different phenomena of nonlinear stiffness and damping taking place in mechanical systems. The *intrawave* frequency *modulation* that indicates nonlinear vibration behavior can be observed in the time domain as modulation. The same phenomenon in the frequency domain can be denoted as energy contributions in the additional spectral peaks. Therefore, *intrawave* frequency *modulation* is depicted in a HT analysis as well as in a Fourier analysis. Stiffening, softening stiffness, dry and turbulent friction are the main typical nonlinear effects illustrated by simple examples. Naturally, stiffness and damping are complex phenomena and more than one of these may exist in the same real structure. Considering the total static force as a sum of typical stiffness and ctional elements, one can write an expression for the total force characteristics as the sum of typical dependencies. However, the decomposition of such total characteristics has no simple solution. It is only when each simple nonlinear mechanism operates at a different range of amplitude and/or forced frequency that we can obtain a unique interpretation of its nonlinear behavior.

9

Identification of the primary solution

By observation (experiment), we acquire knowledge of the position and/or velocity of the object as well as the excitation at several known instants of time. The dynamic properties of the object – including the nonlinearities – can be identi ed by applying varying load levels to the structure and searching for amplitude and frequency dependencies in the system's response. In the case of a free vibration we have only an output signal – the vibration of the oscillators – whereas in the case of a forced vibration we also deal with the input excitation.

Several methods have been developed for the analysis and identi cation of nonlinear dynamic systems and vibration signals (Kerschen *et al.*, 2006). The HT approach to characterize the response of nonlinear vibration systems in the time domain was one such method (Feldman, 1997). The objective was to propose a methodology to identify and classify various types of nonlinearity from measured response data. The proposed methodology concentrates on the HT signal-processing techniques, essentially on extracting a signal envelope and IF, to directly estimate both modal parameters together with the elastic (restoring) and friction (damping) force characteristics. The HT approach presents the system's response as an instant function of time. It is a highly sophisticated method of discovering the existing linear and nonlinear amplitude and frequency dependencies. The HT approach is suitable for any SDOF vibration system and does not require knowledge of the signal or the system parameters. Such nonparametric identi cation will determine not only the amplitude and frequency dependencies, but also the initial nonlinear restoring and damping forces. In modern signal processing, the HT method is more and more widely applied to the analysis and identi cation of nonlinear dynamic structures (Luo, Fang, and Ertas, 2009).

Hilbert Transform Applications in Mechanical Vibration, First Edition. Michael Feldman.
© 2011 John Wiley & Sons, Ltd. Published 2011 by John Wiley & Sons, Ltd.

9.1 Theoretical bases of the Hilbert transform system identification

The HT identi cation in a time domain is based on the analytic signal representation $X(t) = x(t) + j\tilde{x}(t)$, where $\tilde{x}(t)$ is the HT projection of the solution $x(t)$. The method uses envelope and phase signal representation $X(t) = A(t)e^{j\psi(t)}$, where $A(t)$ is an envelope (or magnitude) and $\psi(t)$ is an instantaneous phase; as both are real functions, $x(t) = A(t)\cos\psi(t)$, $\tilde{x}(t) = A(t)\sin\psi(t)$, $A(t) = \sqrt{x^2(t) + \tilde{x}^2(t)}$, $\psi(t) = \arctan[\tilde{x}(t)/x(t)]$. The system solution $x(t)$ is known by direct measurements, and the HT projection $\tilde{x}(t)$ can be computed with the use of a Hilbert transformer (Section 3.3). The envelope, the phase, and their derivatives can be computed directly as functions of time or by the use of the analytic signal relations (Section 4.9): $\dot{X} = X(\frac{\dot{A}}{A} + i\dot{\psi})$, $\ddot{X} = X(\frac{\ddot{A}}{A} - \dot{\psi}^2 + 2i\frac{\dot{A}}{A}\dot{\psi} + i\ddot{\psi})$, where $\dot{\psi}(t) = \omega(t) = \dfrac{x(t)\dot{\tilde{x}}(t) - \dot{x}(t)\tilde{x}(t)}{A^2(t)} = \text{Im}[\dfrac{\dot{X}(t)}{X(t)}]$ is the IF of the solution $x(t)$.

Now, consider the equation of the system motion (8.10), which includes both the restoring and damping forces $\ddot{x} + h(x)\dot{x} + \omega_0^2(x)x = 0$. We already know that a nonlinear restoring force, as a function of time, can be transformed into a multiplication form $\omega_0^2(x)x = \omega_0^2(t)x(t)$ with a new fast-varying natural frequency $\omega_0^2(t)$ and a system solution $x(t)$ with an overlapping spectra. Similarly, the nonlinear damping force can also be transformed into a function of time as a multiplication $h_0(x)\dot{x} = h_0(t)\dot{x}(t)$ between the fast-varying instantaneous damping coef cient $h_0(t)$ and the velocity, with an overlapping spectra. As a result, we will have an instantaneous natural frequency $\omega_0^2(t)$ and instantaneous damping coef cient $h_0(t)$ with highpass overlapping characteristics, with the solution $x(t)$ in the equation of motion:

$$\ddot{x} + 2h_0(t)\dot{x} + \omega_0^2(t)x = 0 \qquad (9.1)$$

To nd the HT of the product of the functions with an overlapping spectra we will use a generalized form of the HT multiplication (Section 2.4). It allows us to produce the HT of the differential equation of the motion without the assumption of nonlinear functions. When the HT of (9.1) is derived, it will pass through the multiplication functions $\omega_0^2(t)x(t)$ and $h_0(t)\dot{x}(t)$. Further, the Hilbert transformation commutes with differentiation, so we will get

$$\ddot{\tilde{x}} + 2\overline{h}_0(t)\dot{\tilde{x}} + 2\tilde{h}_1(t)\dot{x} + \overline{\omega_0^2}(t)\tilde{x} + \widetilde{\omega_1^2}(t)x(t) = 0 \qquad (9.2)$$

Multiplying by i and adding (9.2) to (9.1) yields a differential equation of motion in signal analytic form, that is,

$$\ddot{X} + 2h(t)\dot{X} + \omega_x^2(t)X = 0, \qquad (9.3)$$

where $\omega_x(t)$ is a fast-varying *instantaneous modal frequency*, $h(t)$ is a fast-varying *instantaneous modal damping coefficient*, and $X = x(t) + i\tilde{x}(t)$.

These two instantaneous parameters of Equation (9.3) are unknown; however this complex equation also consists of two separate equations for real and imaginary parts. We have the same number of linear equations as the number of unknown variables, so the system is exactly solvable with only one possible solution for instantaneous modal parameters $\omega_x(t)$ and $h(t)$.

9.2   Free vibration modal characteristics

Consider a SDOF nonlinear vibration system (9.3) under a free vibration regime. A second-order conservative system with a nonlinear restoring force and a non-linear damping force has a solution in an analytic signal form $X = x + i\tilde{x}$, where $x(t) = A(t)\cos\left[\int \omega(t)dt\right]$ is a measured vibration solution. At the rst stage of the identi cation technique, the envelope $A(t)$ and the IF $\omega(t)$ are extracted from the vibration signal on the basis of HT signal processing.

Now, the derivatives $\dot{X}$ and $\ddot{X}$ are known functions of $A(t)$ and $\omega(t)$ (see Section 4.9). Substituting the derivatives in (9.3) yields $X\left[\frac{\ddot{A}}{A} - \omega^2 + \omega_x^2 + 2h\frac{\dot{A}}{A} + i\left(2\frac{\dot{A}}{A}\omega + \dot{\omega} + 2h\omega\right)\right] = 0$

Separating the real and imaginary parts and equating them to zero, in order to nd the *instantaneous (natural) modal parameters*, gives:

$$h(t) = -\dot{A}/A - \dot{\omega}/2\omega$$
$$\omega_x^2(t) = \omega^2 - \ddot{A}/A + 2\dot{A}^2/A^2 + \dot{A}\dot{\omega}/A\omega, \qquad (9.4)$$

where A is an envelope, and ω is an IF of the solution.

Both of the obtained instantaneous modal parameters – the natural frequency $\omega_x^2(t)$ and the damping coef cient $h(t)$ – are functions of the rst and second derivatives of the signal envelope and the IF. The instantaneous modal parameters can be calculated directly at every point of the measured free vibration solution $x(t)$. Algebraically, Equations (9.4) mean that the HT identi cation method uses measured displacement, velocity, and acceleration signals together. Equations (9.4) are rather simple and do not depend on the type of nonlinearity that exists in the structure. When applying this direct method for a transient free vibration, the instantaneous modal parameters are estimated directly. Linking the modal frequency and the envelope, we will obtain a skeleton curve from the HT analysis. In the same way – by linking the modal damping and the envelope – we will get the damping curve. Backbones and damping curves are very helpful and are used as a traditional instrument in nonlinear vibration analysis.

Note, that the instantaneous natural frequency $\omega_x^2(t)$ differs from the IF $\omega(t)$ of the signal, because it depends on variations of the signal envelope and also of the IF. For small and slow nonlinear variations – when the second-order members can be neglected ($\dot{A}^2 = \ddot{A} = \dot{\omega} = \dot{A}\omega = 0$) – Equations (9.4) show that the instantaneous modal frequency of the system will be close to the IF of the solution, while the instantaneous damping coef cient will be equal to the ratio between the envelope and its derivative. It is clear that in the case of nonlinearities causing variations of the envelope and the IF, an instantaneous natural frequency becomes a fast-oscillating varying function. The natural frequency will be equal to the IF constant value only for a linear conservative system when $A = \omega = 0$: $\omega_x = \omega$.

According to (9.4), the instantaneous modal damping coef cient $h(t)$ is a function of two rst derivatives – the envelope A and the IF ω – therefore, considering that only a single envelope variation affects a part of damping function, it is incomplete.

Note that, in the case of a linear restoring force and a constant frequency of the solution, the instantaneous damping coef cient $h(t)$ depends only on variations in the envelope (9.4). Let us now determine now an average damping coef cient $\langle h \rangle$ during some time, counting a single full period of the solution or a number of periods. Functions of the form $F(x)/F(x)$ allow a simple integration, which produces a natural logarithm $\ln |F(x)|$. The derivative of a natural logarithm with a generalized functional argument $F(x)$ is $\dfrac{d}{dx} [\ln F(x)] = F(x)/F(x)$.

In this way the average damping coef cient gets the form $\langle h \rangle = (t_{i+n} - t_i)^{-1}$ $\int_{t_{i+n}}^{t_i} \dfrac{A(t)}{A(t)} dt = (t_{i+n} - t_i)^{-1} \ln \dfrac{A_i}{A_{i+n}}$. The averaging time can include n number of full periods of the solution $t_{i+n} - t_i = Tn = n/f$, where T is a period and f is a cycle frequency of the solution. As a result we can write $\langle h \rangle = \ln (A_i/A_{i+n}) f/n$. A natural logarithm of the envelope ratio classically determines the logarithmic decrement (8.12) $\ln (A_i/A_{i+n}) = \delta n$. So, nally, an average damping coef cient is found to be equal to the logarithmic decrement multiplied by the solution frequency:

$$\langle h \rangle = \delta f = \delta \omega / 2\pi \qquad (9.5)$$

where $\langle h \rangle$ is the average damping coef cient, δ is the logarithmic decrement, T is the period, and ω is the angular frequency of the damped solution. So a parameter that is well known as the logarithmic decrement measured from the decay envelope is just an average estimation of the damping properties of the vibration system!

Methods of identifying damping in the time domain, that are based on the slopes of the envelope and the logarithmic-decrement method, deal only with an estimate of the integrated damping (Agneni and Balis-Crema, 1989; Yang, Kagoo, and Lei, 2000; Giorgetta, Gobbi and Mastinu, 2007). To produce an oscillating decay motion a tested vibration system is certain to be a lightly damped system with an underdamped term.

This HT approach was used in practice to identify skeleton curves under the free vibration of the following typical nonlinear vibration models: backlash, saturation, precompressed, bilinear, rigid boundary, and Coulomb friction (Wang et al., 2003).

9.3   Forced vibration modal characteristics

A forced vibration regime – unlike free vibration with only nonzero initial conditions – signi es the presence of a continuously acting excitation that generates a forced vibration. We will refer to the same equation of a SDOF vibration system in the analytic signal form (9.3), but with the acting force $Z(t)$ in the right-hand side:

$$\ddot{X} + 2h(t)X + \omega_x^2(t)X = Z/m \qquad (9.6)$$

where $X = x(t) + i\tilde{x}(t) = A(t)e^{i\phi(t)}$ is a complex forced solution of the system, $\omega_x(t)$ is the instantaneous natural frequency, $h(t)$ is the instantaneous damping coefficient, $Z(t)$ is a forced excitation in the analytical signal form, and m is the *modal mass* of the system. Assume that a forced excitation is a monocomponent quasiharmonic signal with a slow-varying (modulated) frequency. Such a harmonic signal, whose frequency sweeps slowly in time, is extremely useful as a stimulus signal for measuring and estimating a system's frequency response function.

Using the system solution X and its derivatives X, $\ddot{X}$ in an analytic signal form we obtain a complex expression for the forced vibration $X\frac{\ddot{A}}{A} - \omega^2 + \omega_x^2 + \frac{2hA}{A} + i(\frac{2A\omega}{A} + \omega + 2h\omega) = \frac{Z}{Xm}$, where A, ω are an envelope and an IF of the solution. Rewriting the last expression as two separate equations for the real and imaginary parts, we receive instantaneous modal parameters as functions of the first and second derivatives of the signal envelope and the IF:

$$\omega_x^2(t) = \omega^2 + \alpha(t)/m + \beta(t)A/A\omega m - \ddot{A}/A + 2A^2/A^2 + A\omega/A\omega$$
$$h(t) = \beta(t)/2\omega m - A/A - \omega/2\omega \tag{9.7}$$

where $\omega_x(t)$ is the instantaneous modal frequency, $h(t)$ is the instantaneous modal damping coefficient, ω, A are the IF and envelope of the vibration with their first and second derivatives $(A, \ddot{A}, \omega)$, and $\alpha(t) = \text{Re}(Z/X)$, $\beta(t) = \text{Im}(Z/X)$ are the real and imaginary parts of the input–output signal ratio. These real and imaginary parts of the input–output signal ratio $Z/X = \alpha(t) + i\beta(t)$ are calculated according to $\alpha(t) = \dfrac{zx + \tilde{z}\tilde{x}}{x^2 + \tilde{x}^2}$ and $\beta(t) = \dfrac{\tilde{z}x - z\tilde{x}}{x^2 + \tilde{x}^2}$.

Making a comparison between Equations (9.7) and (9.4), we can see that Equation (9.7) for the instantaneous modal parameters in the case of forced vibration is more general, because – in addition to members with an envelope, IF, and their derivatives – it includes members with input–output signal ratios. When there is no excitation of the system $(Z = 0)$, Equation (9.7) becomes equal to Equation (9.4) for the instantaneous modal parameters of a free vibration. The input–output signal ratio in (9.7) incorporates both the steady state and the transient part of the solution. The presence of the first and second derivatives of the signal envelope and the IF in (9.7) reflects the transient effect of the solution and determines the modal parameters in more complicated testing conditions, for instance, when the excitation is a nonstationary quasiharmonic signal with a high sweep frequency.

In the case of a forced vibration, we have three unknown modal parameters: the frequency, damping, and mass, while the obtained set consists only of two separate equations for the real and imaginary parts (9.7). Thus, because of the *a priori* unknown mass we cannot calculate all the instantaneous modal parameters directly. To estimate the modal parameters we will use the following minimization approach. Let us assume that the modal parameters of the system under a forced vibration regime do not deviate widely in most practical cases. The best available value of the reduced mass is a value that minimizes the deviation of the instantaneous modal frequency and the modal damping coefficient in (9.7) during specific time segments. The optimal reduced mass value will be the modal effective inertia mass of the SDOF vibration system. As we have the reduced the modal mass value, we can calculate the instantaneous modal frequency and the instantaneous modal damping coef cient at every time point of the solution – directly, according to (9.7).

In terms of engineering, an input generally is some measured signal $z(t)$, and an output is another measured signal $x(t)$. Thus, using the vibration output and excitation input signals, their Hilbert projections, and the rst and second derivatives of the vibration, we can determine the instantaneous modal parameters of the system at every point of the solution. The resulting nonlinear algebraic equations (9.7) are fairly simple and do not depend on the type of nonlinearity that exists in the structure. It is essential that the frequency of the input excitation has to vary slowly in time, exciting a forced vibration with different frequencies.

It is well known that, during a fast swept frequency excitation, all resonances excited in the structure have a lower and frequency-shifted maximum response than the maximum response expected for a steady-state harmonic excitation with a similar resonance frequency. Nevertheless, the modal parameters estimated according to the HT force identi cation method do not depend on the rate of the sweeping frequency and are the same as for the steady-state resonance harmonic excitation.

The obtained instantaneous modal parameters are varying functions of time and have a fast oscillation around their smooth average values (Davies and Hammond, 1987; Gottlieb and Feldman, 1997). The fast oscillation occurs because the instantaneous modal parameters represent not only the average system values, but also fast intramodulations during any short part of the solution period.

9.4   Backbone (skeleton curve)

In most cases, if a conservative system to be analyzed has nonlinear elastic forces, the frequency of a free vibration will decisively depend on the amplitude of vibrations. The traditional theoretical backbone of a nonlinear system is a dependency between (a) the natural frequency of free vibration corresponding to a single full vibrational cycle and (b) the displacement amplitude. The representation of a traditional vibration system indicates that a conservative nonlinear system has a constant amplitude and a constant natural frequency during every full period of oscillation. This means, for example, that the backbone of a conservative Duf ng system should be mapped just as a single point with a constant amplitude and frequency. If damping that is responsible for decreasing the amplitude is present in a nonlinear system, the traditional backbone transforms from a point to a smooth curve. Some typical nonlinear examples of a backbone representation (Figure 9.1) were considered in Section 8.3. In a linear system the stiffness characteristics do not change, so the skeleton curve will be a trivial straight line.

A backbone is a very helpful and traditional instrument in vibration analysis; therefore it is conventional to use the HT approach to construct backbones. The HT analysis can directly present a skeleton curve (backbone) as a function between the envelope and the modal frequency. However, the obtained instantaneous modal frequency $\omega_x(t)$ and envelope $A(t)$ include both slow-varying (constant for a linear system) and fast, intracycle varying components. As a result, the HT backbone with the presence of a nonlinearity, will not be a smooth curve, but a smeared-out

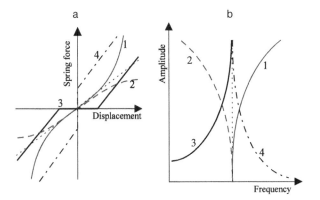

Figure 9.1 Typical nonlinear stiffness force characteristics (a) and backbones (b): hardening (1), softening (2), backlash (3), preloaded (4) (Feldman, ©2009 by John Wiley & Sons, Ltd.)

(oscillated) one; the HT backbone of a conservative Duf ng system will be not a point, but a short-length tilted line (Figure 8.13). To utilize the HT analysis in a traditional smooth backbone presentation we should try to smooth or to decompose the instantaneous fast intracycle varying modal parameters.

However, the HT identi cation based on a direct time domain estimation of the modal parameters allows us to extract and plot the backbone curve as a dependency between the envelope and the instantaneous modal frequency of the system. Note, that there are no assumptions on the forms of $A(\omega_x)$, so the HT identi cation method is truly nonparametric.

9.5   Damping curve

In the same way that a backbone depicts a modal frequency as an amplitude function, a damping curve draws a modal damping coef cient as an amplitude function. A damping curve is an inherent feature of nonlinear systems showing that the damping system property is amplitude dependent. In linear systems the damping characteristics do not change, so the damping curve will be a trivial straight line. In a nonlinear case, when the modal damping depends on the vibration frequency, the damping curve depicts the modal damping coef cient as a function of the frequency. Some typical nonlinear examples of a damping curve representation (Figure 9.2) were considered in Section 8.1. Note, that there are no assumptions on the forms of the damping curve $A(h)$, so the method is truly nonparametric.

9.6   Frequency response

As a result of the HT identi cation, a set of duplet modal parameters (the instantaneous modal frequency ω_x and the instantaneous modal damping h of each natural mode of vibration) is de ned. In practice it is convenient to present the identi ed modal

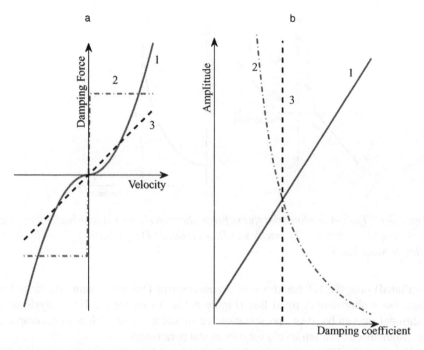

Figure 9.2   Typical nonlinear damping force characteristics (a) and damping curves (b): turbulent (1), dry (2), viscous friction (3)

parameters in the traditional form of the FRF in a speci ed frequency range using a harmonic excitation as an input to the vibration system at every frequency (Feldman, 1997). Thus the tested system harmonic response amplitude expressed as a linear SDOF can be written as:

$$A = \frac{2A_{max}h(A)}{\omega_x(A)\sqrt{\left[1 - \dfrac{\omega^2}{\omega_x^2(A)}\right]^2 + \dfrac{4h^2(A)\omega^2}{\omega_x^2(A)}}} \tag{9.8}$$

Here A is the steady-state vibration amplitude (proportional to the FRF magnitude), A_{max} is the maximum (resonance) value of the vibration amplitude, $\omega = 2\pi f$ is the angular frequency of the vibration, $\omega_x(A) = 2\pi f_x(A)$ is the angular modal frequency as a function of the amplitude, and $h(A)$ is a damping coef cient as a function of the amplitude. For the sake of plotting the estimated response of the tested system using the identi ed modal parameters, Equation (9.8) should be further inverted:

$$\omega^2 = \omega_x^2(A) - 2h^2(A) \pm 2\omega_x(A)h(A)\sqrt{\frac{A_{max}^2}{A^2} - 1 + \frac{h^2(A)}{\omega_x^2(A)}},$$

$$0 \leq A \leq A_{max}\left[1 - \frac{h^2(A)}{\omega_0^2(A)}\right]^{-\frac{1}{2}} \tag{9.9}$$

Using (9.9) we can plot the tested system harmonic response amplitude in a common form as a separate resonance curve and the system backbone curve. If the identi ed modal parameters are amplitude dependent functions, the FRF obtained will re ect these nonlinearities in the left and right branches and the bandwidth between them.

A similar concept, known as the Nonlinear Output Frequency Response Function, was proposed in Lang and Billings (2005). The concept can be considered to be an alternative extension of the classical FRF for linear systems of the nonlinear case. This extension is a one-dimensional function of frequency, which allows the analysis of nonlinear systems to be implemented in a manner similar to the analysis of linear systems, and provides great insight into the mechanisms that dominate in nonlinear behavior.

9.7   Force static characteristics

The time-domain identi cation based on the HT allows for the direct extraction of the system's linear and nonlinear instantaneous modal parameters. Thus we can consider the concluding stage of an inverse problem, namely, the identi cation of the initial average nonlinear elastic and damping static force characteristics from the equation of a vibration motion. According to formulas (9.4) and (9.7), we are able to get a duplet of instantaneous modal parameters, such as an instantaneous modal frequency ω_x and an instantaneous modal damping h. In the next stage, the symmetric nonlinear elastic and damping force characteristics can be estimated trivially according to the following expression:

$$k(x) \approx \begin{cases} \omega_x^2(t)A(t), & x > 0 \\ -\omega_x^2(t)A(t), & x < 0 \end{cases}; \quad h(x)x \approx \begin{cases} h(t)A_x(t), & x > 0 \\ -h(t)A_x(t), & x < 0 \end{cases} \quad (9.10)$$

where $A(t)$ and $A_x(t)$ are envelopes of the displacement and the velocity of the vibration motion, respectively, $k(x)$ is the restoring force per unit mass as a function of the displacement x, and $h(x)x$ is the damping force per unit mass as a function of the velocity x. Each obtained force characteristic is a relative characteristic, dealing only with a unit mass of the vibration system, and each obtained force characteristic is considered to be a static symmetric characteristic. The nal estimation of the force characteristics is based only on modal parameters and does not require intermediate backbones or FRFs.

As mentioned, each instantaneous modal parameter as a function of time varies fast around its average value. These fast oscillations were shown, for example, in a the parameter estimation of mooring systems using the HT in Gottlieb, Feldman, and Yim (1996). The fast uctuations are the result of two main factors. First, a measured nonlinear vibration is a composition of a primary solution and an in nitive number of high ultraharmonics. Second, the estimated instantaneous modal parameters re ect the presence of ctitious nonlinear members of the stiffness in a damping and of the damping in a stiffness. This fast oscillating appearance of elastic and damping forces contradicts the classic smooth form of static force characteristics used in the initial differential equation of motion. To get back the smooth initial static force characteristics we can use several techniques.

9.7.1   Averaging of the instantaneous modal parameters

The simplest way of removing fast oscillations is by a lowpass ltering of the instantaneous modal parameters: $A(t), \omega_x(t), h(t)$. As a result, we will have only slow-varying modal parameters $\langle A(t) \rangle$, $\langle \omega_x(t) \rangle$, $\langle h(t) \rangle$ smoothing the restoring and damping forces – according to (9.10). A subsequent lowpass frequency ltering of the instantaneous modal parameters makes it possible to construct smooth nonlinear backbone curves for both restoring and damping functions (Feldman, 1991). However, as will be explained in Chapter 11, these averaged modal parameters and static force characteristics are biased against the true initial nonlinear characteristics.

9.7.2   Polynomial scaling technique

It was shown in Section 8.6 that the average natural frequency simply repeats the structure of the initial nonlinear elastic characteristics correct to the corresponding polynomial numeric coef cients. By constructing a polynomial curve tting of the dependency between the average natural frequency and the envelope, we can estimate these polynomial coef cients. The reconstructed smooth initial nonlinear force characteristics will have the same form of curve tting but with recalculated coef cients according to (8.9). Such a scaling technique is better than a simple averaging, but is not universal because it can only be applied to polynomial-type nonlinearities.

9.7.3   Selecting extrema and scaling technique

It is known that the total energy of a conservative vibration system is constant, and during the free vibration in each moment the energy is partly kinetic and partly potential. To estimate each force function more precisely we can utilize the fact that the envelope value is equal to the peak value of the displacement around every maximum displacement point. Moreover, at each maximum displacement point the corresponding value of velocity is close to zero, so its contribution to the varying natural frequency is negligibly small. In the same way, around every peak point of the velocity, the corresponding displacement value is close to zero, so its contribution in the varying damping coef cient is negligibly small. This means that multiplying the instantaneous modal frequency squared and the envelope at the point of maximum displacement $t_{x\,max}$ gives the exact value of the stiffness static force value $k(x_{max}) = \omega_x^2(t_{x\,max})A(t_{x\,max})$. Respectively, multiplying the instantaneous modal damping and the velocity envelope at the point of maximum velocity $t_{\dot{x}_{max}}$ gives the exact value of the friction static force $h(x_{max})x_{max} = h(t_{\dot{x}\,max})A_{\dot{x}}(t_{\dot{x}\,max})$. The number of these peak points is far less than the total number of points in a vibration signal. Therefore we could recommend, rst, averaging the envelope and the modal frequency as biased functions $\langle \omega_x^2(t) \rangle$, $\langle h(t) \rangle$, where the angle bracket sign indicates the averaging procedure; and, second, the calculating the scale factor functions s_ω, s_h. That will adjust the full length average functions to their unbiased values estimated around only the extrema points $s_\omega = \omega_x^2(t_{x\,max})/\langle\omega_x^2(t)\rangle$, $s_h = h(t_{x\,max})/\langle h(t)\rangle$ (Feldman, 1997). Finally, we will calculate the unbiased elastic force as a function of the displacement and a damping force as a function of the velocity:

$$
k(x) \approx \begin{cases} s_\omega \langle\omega_x^2(t)\rangle \langle A(t)\rangle\,, & x > 0 \\ -s_\omega \langle\omega_x^2(t)\rangle \langle A(t)\rangle\,, & x < 0 \end{cases};\quad h(x)x \approx \begin{cases} s_h \langle h(t)\rangle \langle A_x(t)\rangle\,, & x > 0 \\ -s_h \langle h(t)\rangle \langle A_x(t)\rangle\,, & x < 0 \end{cases}.
$$

The computational procedure is rather simple. First, we extract fast-varying instantaneous parameters via the HT. Then we eliminate fast oscillations by lowpass ltering and shift smooth parameters by scaling. Finally, we estimate the smooth nonlinear elastic and friction force characteristics. The described technique is realized in the FREEVIB and FORCEVIB identi cation methods (Sections 10.5.2–10.5.3).

9.7.4   Decomposition technique

All the previously described techniques to identify the force just eliminate the fast oscillation and operate only with the slow parts of the instantaneous modal parameters. As a result we have a rather good but only approximate estimation of the force characteristics. Instead of removing fast oscillations from the instantaneous modal parameters, we can just exploit them completely. The idea of using fast-varying modal parameters is based on the HT decomposition of a varying nonstationary signal (Feldman, 2006). Using all existing high-frequency signal components allows us to identify the smooth force characteristics directly and precisely. The details of the precise HT identi cation will be discussed further in Chapter 11.

9.8   Conclusions

We can draw the following conclusions from the representation of an analytical signal. Both the IF and the amplitude of free vibration are complicated modulated signals. Nonlinear solutions can be represented by an expansion of members with different frequencies, or by a time-varying signal with ab oscillated instantaneous frequency and envelope. The instantaneous frequency and envelope of a nonlinear vibration obtained via the HT are time-varying fast-oscillating functions. For example, in the presence of a cubic nonlinearity and threefold high harmonics, the frequency of the instantaneous parameter oscillation is twice that of the main frequency of vibration. The dependency between the average envelope and the average instantaneous frequency plots a backbone that is close to the smooth theoretical backbone of nonlinear vibration. Using the proposed HT analysis in the time domain, we can extract both the instantaneous undamped frequency and the average nonlinear elastic force characteristics.

As nonlinear dissipative and elastic forces have totally different effects on free vibration (energy dissipation lowers the instantaneous amplitude, while nonlinear elasticity links the instantaneous frequency and amplitude in a certain relationship), it is possible to determine some aspects of the behavior of these forces. For this identi cation we propose that relationships be constructed between the instantaneous frequency and amplitude plus curves of the instantaneous decrement as a function of amplitude. The identi cation technique developed here should be of value in many areas of mechanical oscillatory systems that have various features of a nonlinear behavior.

10

The FREEVIB and FORCEVIB methods

The nonparametric identification of nonlinear vibrating oscillators, being a typical dynamics inverse problem, deals with the construction of initially unknown functions of nonlinear restoring and damping forces. Typically, every nonlinear equation describing a vibration motion has a fixed structure. This structure classically includes three independent members: a restoring elastic force (stiffness, spring) as a nonlinear function of a displacement (position), a damping force (friction) as a nonlinear function of velocity (the first derivative of a position with respect to time), and an inertial force proportional to acceleration (the second derivative of a position with respect to time). Every independent restoring and damping force member is an *a priori* unknown nonlinear function of motion. For example, a hardening or softening spring, dry or turbulent friction, must be characterized by a function of the state rather than by a single scalar parameter. By observation (experiment), one acquires knowledge of the position and/or velocity of the vibrating object as well as the excitation at several known instants of time.

The described nonparametric identification determines the initial nonlinear restoring and damping forces. In the case of a free vibration, one can acquire only a response signal – the vibration of the oscillators – whereas in the case of a forced vibration one can utilize both the measured excitation and the response. The HT time-domain identification is an effective instrument for the evaluation of the particular properties of a nonlinear vibration system – for example, for the reconstruction of the characteristics of elastic and dissipative nonlinear forces. For this identification we propose that relationships be reconstructed between the instantaneous modal frequency and displacement, together with curves of the instantaneous damping coefficient as a function of velocity. The described method of analyzing the vibration of a machine offers a way to the direct plotting of a skeleton and damping curve for the system; this contains the values of instantaneous modal parameters, and the spring and

Hilbert Transform Applications in Mechanical Vibration, First Edition. Michael Feldman.
© 2011 John Wiley & Sons, Ltd. Published 2011 by John Wiley & Sons, Ltd.

friction force characteristics. The identification technique presented here should be of value in the investigation of many areas of mechanical oscillatory systems, including aspects of their nonlinear behavior.

The time-domain HT approach had some success as a direct method of non-parametric identification. There are essentially two approaches, one based on a free vibration FREEVIB (Feldman, 1994a) and the other on a forced vibration FORCE-VIB (Feldman, 1994b). Both approaches provide a method of obtaining the stiffness and damping characteristics of SDOF systems. Both approaches separately identify the system modal parameters; and even stiffness and damping nonlinearities appear in combination. They both take into account only the solution of the primary system, ignoring all other high-frequency ultraharmonics. The FORCEVIB also identifies the inertia modal parameter, while FREEVIB allows us to identify only the elastic and damping modal parameters. In general these HT identification methods include the following sequence of operations:

- Taking the HT of the measured vibration (and excitation) signals and calculating their envelope and the IF.

- Calculating the instantaneous modal parameters – such as the modal frequency, modal damping, modal mass value.

- Lowpass filtering of the modal parameters; calculating the scale factor functions around the selected extrema points of displacement and velocity; scaling the smooth modal parameters.

- Presenting the results in the form of the backbones, the damping curves, the FRF, and the force static characteristics.

The resultant backbones – because of their strong association with nonlinearities – not only detect the presence, but also show up most vividly the specific type and level of existing nonlinearities in the vibration system. The final force static characteristics reveal the initial nonlinear vibration model. Such a final construction of the identified model does not even require knowledge of a certain association between nonlinearities and backbones. The final force static characteristics are estimated automatically and directly by the FREEVIB and FORCEVIB methods.

A vibration signal, suitable for this identification, should be a monocomponent signal derived directly from a SDOF system or obtained from a MDOF system after a signal decomposition or after a bandpass filtering. The simplest practical way to get a monocomponent free vibration signal even from a MDOF system is to first excite a forced vibration around the desirable resonance frequency, and suddenly turn off the excitation. The remaining free decay will be a free solution of the desirable natural mode of the nonlinear vibration system. This output signal $x(t) = A(t) \cos \int \omega(t) dt$, where $x(t)$ is the vibration signal (a real-valued function), $A(t)$ is the envelope (an instantaneous amplitude), and $\omega(t)$ is the IF as a monocomponent signal, is suitable for a FREEVIB analysis. The analytic signal representations permit us to estimate the instantaneous modal frequency, the instantaneous damping coefficient, and the static force characteristics, and to consider the equivalent equation of motion as $\ddot{x} + 2h(\dot{x})\dot{x} + k(x) = 0$.

Algebraically, FREEVIB and FORCEVIB equations mean that the HT identification methods use the initial displacement, velocity, and acceleration all together and synchronously. Note however that, in practice, the necessity of the first- and second-order derivatives will increase the noise level in the estimations because of precision errors inherent in the digital signal processing. Thus, the HT identification methods, being sensitive to noise, require high-quality experimental measurements with a minimum of instrumental noise and random errors.

Some other existing techniques for the identification of the nonlinear stiffness and damping in a mechanical structure assume an *a priori* known particular type of stiffness and damping model. After taking measurements they try to fit the model parameters with regard to some specific structure. Such a selection of the chosen model hides the physics of the system.

The HT identification methods, as nonparametric systems, form the acting non-linear elastic and damping force characteristics $k(x)$, $h(\dot{x})$ by directly extracting the vibration system modal parameters. A nonlinear spring force function, identified from the SDOF system's motion of vibration, will totally correspond to the system's initial static elastic force characteristics per unit mass. In general, a SDOF system could include several elastic and damping individual elements, combined integrally through parallel and/or series connections. If each individual element of a SDOF system is known, it is usually possible to determine the resultant, or equivalent, system force characteristics. But the inverse problem has no unique solution.

For example, consider a system with two spring elements connected in parallel, where the first is a backlash with a clearance and the second is just a linear spring. The corresponding equivalent force characteristics will have a bilinear form with a linear section for the displacement, less than the clearance, and with another linear section for a higher amplitude. If the data on equivalent force characteristics is the only information available, it is not possible to reconstruct the initial system. In our example it could be either a real bilinear element, or just two different initial elements like a backlash or a linear spring. In complicated cases of the system identification, there are no unique solutions for the decomposition of the resultant force characteristics, and one should use some additional information of the structure of the model and a combination of its elements.

The HT methods, namely FREEVIB and FORCEVIB, of free and forced vibration analysis determine instantaneous modal parameters, even if the input signal is a high-sweep frequency signal. Such a direct determination of the relationship between an amplitude and a natural frequency, which characterizes the elastic properties, and a relationship between an amplitude and the damping characteristics, makes an efficient nonlinear system testing possible without a long forced response analysis. The reduction of time required for dynamic testing is the one of important advantages of the FREEVIB and FORCEVIB methods. The methods consider the steady state as well as the transient relations between the input and output signals of a vibration system. Therefore, for the forced vibration testing with sweeping frequency there is no need to wait until the end of the transient regime.

Note that the nonlinear backbones obtained according to FORCEVIB have two close branches forming a narrow loop instead of a single-valued curve. The branches usually correspond to the input frequency before and after passing through the system resonance. For a smaller nonlinearity and slower sweeping frequency, the loop will tighten to a single-valued backbone curve. In a nonlinear system like the Duffing oscillator, there can be two types of stationary responses with jumps as transitions from one response to another. The transition over resonance corresponds to the switching from different jump-up to jump-down resonance frequencies that occur under forced vibration (Brennan *et al.*, 2008).

In the case of linear systems the FORCEVIB method is able to estimate precisely the modal parameters for very fast sweeping, including only a few full cycles of the sweeping excitation around the resonance frequency. A forced vibration testing of nonlinear systems requires more time and a slower sweeping frequency because of the appearance of ultrahigh nonlinear harmonics. Nevertheless the HT identification methods are less time consuming than traditional spectral analysis techniques.

10.1   FREEVIB identification examples

The FREEVIB method is illustrated here using the data from numerical simulation results published on the internet as the MATLAB® pre-parsed pseudo-code files (P-file) (Feldman, 2010). The first example is a free vibration of the Duffing equation $\ddot{x} + 2 \times 2.5\dot{x} + (15 \times 2\pi)^2 x + 2000x^3 = 0$; $x_0 = 8$. The subsequent response of the system will take the form of a monotonically decreasing free decay (Figure 10.1).

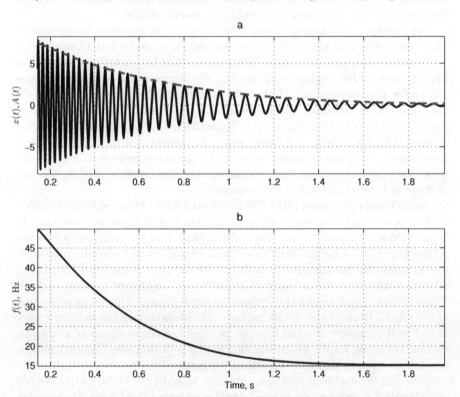

Figure 10.1   The free vibration of the Duffing equation: the displacement (a,—) and the envelope (a, - -) of the solution, the IF of the solution (b) (Feldman, ©2011 by Elsevier)

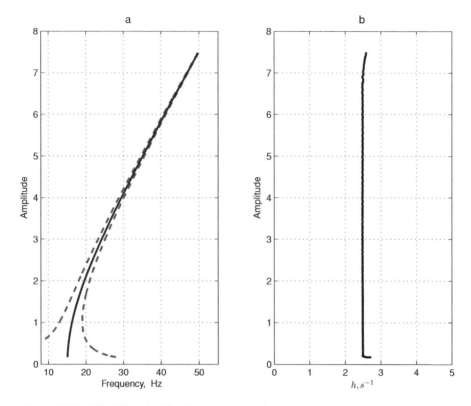

*Figure 10.2   The identified Duffing equation: the backbone (a, —), the FRF (a, --);
the damping curve (b) (Feldman, ©2011 by Elsevier)*

For large amplitudes the estimated backbone and the FRF are deviated to the
right, showing a hardening type of nonlinear stiffness (Figure 10.2a). Small amplitude values have almost no influence on the modal frequency of the system's linear
part, which is equal to 15 Hz. The obtained damping curve, being a trivial vertical
line, just illustrates a linear value of the constant damping coefficient equal to 2.5 s⁻1
(Figure 10.2b). The identified restoring force static characteristic has a typical cubic
form (Figure 10.3a). The estimated function almost agrees with the initial restoring force function $k(x) = (15 \times 2\pi)^2 x + 2000x^3$ plotted as a dash-dot line on the
same figure. The estimated damping force characteristic (Figure 10.3b) is a straight
line following the linear damping part of the equation $2h(\dot{x})\dot{x} = 2 \times 2.5\dot{x}$.

For the next example we examined a free vibration of the system with combined
nonlinearities both in the restoring and in the damping forces: $\ddot{x} + 1100 \, \text{sign}(\dot{x}) +$

$$k(x) = 0, \quad x_0 = 10, \quad \text{where } k(x) = \begin{cases} 0, & \text{if } |x| \leq 1.5 \\ (2\pi \times 30)^2 \, (|x| - 1.5) \, \text{sign}(x), & \text{if } |x| > 1.5 \end{cases},$$

is a nonlinear backlash force, and $1100 \, \text{sign}(\dot{x})$ is a dry friction force. The free
vibration signal and the envelope function are shown in Figure 10.4a. The system
oscillates with a slightly varying frequency and an amplitude that gradually decreases
to zero.

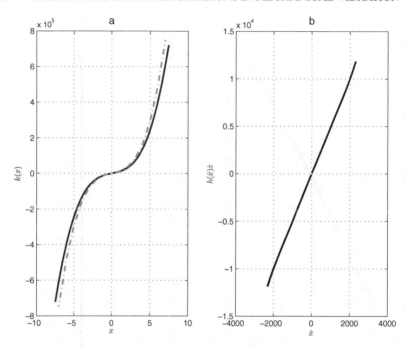

Figure 10.3   The identified Duffing equation spring force (a, —), the initial spring force (a, --); the damping force (b) (Feldman, ©2011 by Elsevier)

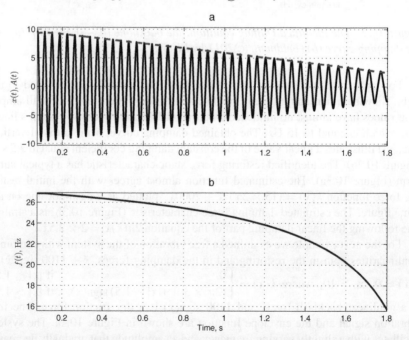

Figure 10.4   Free vibration of the system with a backlash and a dry friction: the displacement (a, —), the envelope (a, --), the IF of the solution (b)

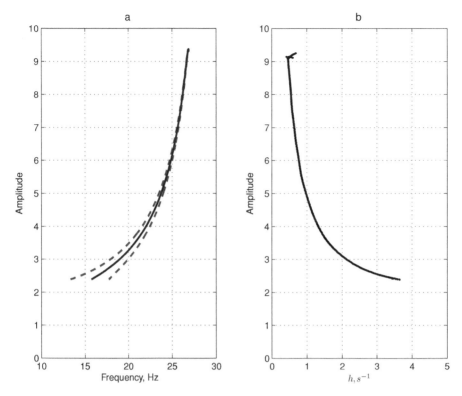

Figure 10.5   The identified system with the backlash and dry friction: the backbone (a, —) and the FRF (a, - -); the damping curve (b)

The backlash system will display its nonlinear properties mainly for small vibrations in amplitude where the natural frequency decreases extensively with decreasing amplitude (Figure 10.4b). The backbone of the backlash system is a monotonically increasing curve (Figure 10.5a) with an asymptote on the right where the natural frequency of the corresponding linear system without the backlash is constant and equal to 30 Hz. The estimated backbone also cuts off a clearance value equal to 1.5 on the amplitude axis on the left side. Due to the dry friction, the damping curve takes the form of a monotonic hyperbola (Figure 10.5b). The calculated force characteristics (Figure 10.6) show good enough agreement with the original restoring and friction forces of the vibration system plotted as a dash-dot line in the same figure.

Note that each force function is obtained automatically without any assumption of the forms of nonlinearity, so the identification method is truly nonparametric.

10.2   FORCEVIB identification examples

To illustrate the identification results of nonlinear systems under a forced vibration we will take the same two dynamics models. The first model is the Duffing equation

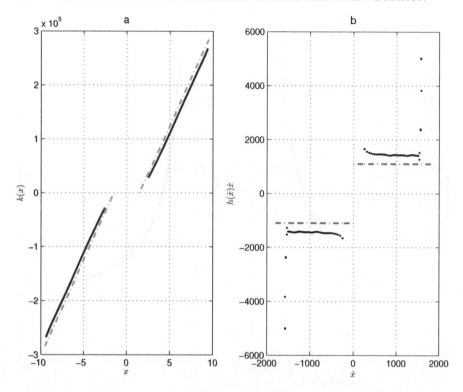

Figure 10.6   The backlash and dry friction system identified spring force (a, —), the initial spring force (a, --), the identified damping force (b, —), the initial damping force (b, ---)

$\ddot{x} + 2 \times 2.5\dot{x} + (30 \times 2\pi)^2 x + 2000x^3 = z$ with a unit mass and an external quasi-harmonic force excitation having a constant amplitude of 1 to 3 and an increasing swept frequency from 20 to 70 Hz over a 2 second period. The measured forced vibration solution is shown in Figure 10.7a with the envelope function. The IF of the swept excitation, the instantaneous modal frequency, and the phase shift between input and output signals are also shown in Figure 10.7b.

The modal parameters obtained by the FORCEVIB method in the form of the backbone curve, the FRF, and the damping curve are presented in Figure 10.8; the estimated mass value of 0.994 is almost equal to the initial value of 1. The identified static force characteristics (with the initial elastic and damping force characteristics as dash-dot lines) are shown in Figure 10.9. The forced vibration regime and measurements lasted only 2 seconds. The natural frequency of the linear member being 30 Hz means that only 60 full vibration cycles were used for the identification procedure. Such a rapid sweep generates high-order ultraharmonics that are ignored by the FORCEVIB method, so the estimated nonlinear static elastics force characteristics slightly differ from the initial characteristics (Figure 10.9a). To increase the accuracy

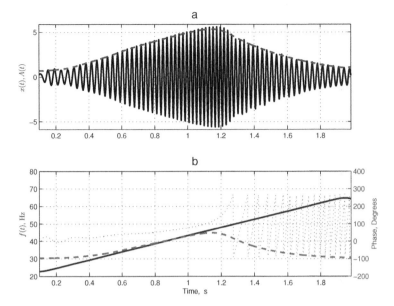

*Figure 10.7   The forced vibration of the Duffing equation: the displacement (a,—)
and the envelope (a, - -); the IF of the swept excitation (b, —), the instantaneous
modal frequency (b, - -), the phase shift between an input and an output (b,...)*

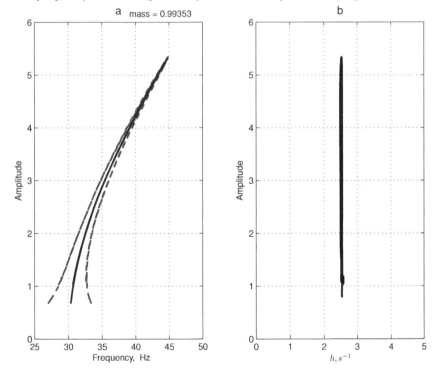

*Figure 10.8   The identified Duffing equation: the backbone (a, —), the FRF (a, - -);
the damping curve (b)*

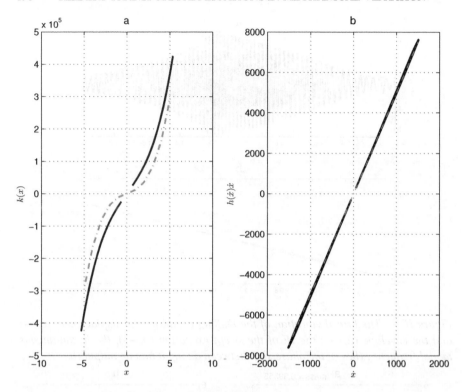

Figure 10.9   The Duffing equation identified spring force (a, —), the initial spring force (a, - -); the damping force (b)

of the nonlinear modal parameters estimation, we need to slow down the sweeping velocity and use longer forced vibration regimes for the dynamic testing.

In the next example we will discuss a system with combined backlash and dry friction nonlinearities (Figure 10.10) $\ddot{x} + 1500\ \text{sign}(\dot{x}) + k(x) = z$, where
$$k(x) = \begin{cases} 0, & \text{if } |x| \le 1.5 \\ (2\pi \times 30)^2 (|x| - 1.5) \, \text{sign}(x), & \text{if } |x| > 1.5 \end{cases}$$ is the nonlinear backlash force, $1500\ \text{sign}(\dot{x})$ is the dry friction force, and z is an external quasiharmonic force excitation of the constant amplitude 13,000 and of the swept frequency increasing from 10 to 70 Hz during a 2-second period. The obtained modal parameters are shown in Figure 10.11 where the backbone skews left and the damping curve skews right towards the lower amplitude. The estimated mass value of 1.05 differs from the initial value by only 5%. Figure 10.12 presents the identified static force characteristics for both the elastic and damping members of the equation, together with the initial force characteristics plotted with dash-dot lines. As can be seen, the nonlinear forces identified from the measured vibrations of the SDOF system correspond well to the system's initial static elastic force characteristics.

It is important that not only is the presence of a nonlinearity detected, but also that an adequate and readily interpretable dynamic system model is identified. The HT

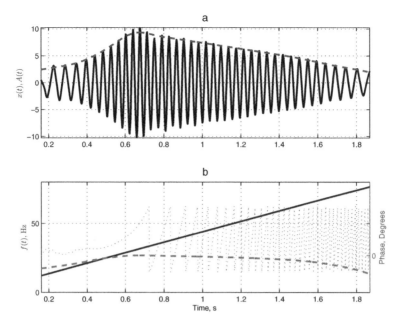

*Figure 10.10   The forced vibration of the system with the backlash and dry friction:
the displacement (a, —), the envelope (a, - -); the IF of the swept excitation (b,—),
the instantaneous modal frequency (b, - -), the phase shift between an input and an
output (b , . . .)*

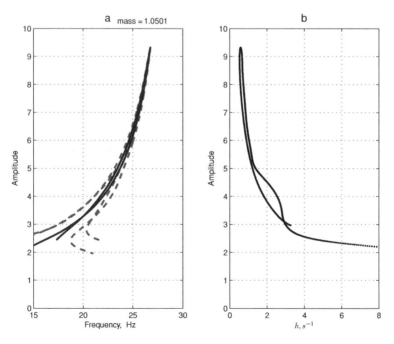

*Figure 10.11   The identified system with the backlash and dry friction: the backbone
(a, —) and the FRF (a, - -); the damping curve (b)*

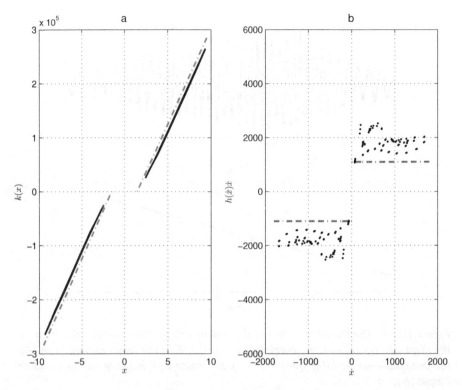

Figure 10.12 The backlash and dry friction identified spring force (a, —), the initial spring force (a, - -); the identified damping force (b,..., the initial damping force (b, ---)

reconstruction of the characteristics of the nonlinear forces is a direct and automatic identification that does not require supplementary analysis of the topography of the skeleton or damping curves. This topography is not essential for an experimental evaluation of the properties of an *a priori* unknown vibration system. Using the FREEVIB and FORCEVIB approaches to identify the SDOF vibration system we can obtain both the modal parameters and nonlinear force characteristics during free or forced vibration analysis.

10.3   System identification with biharmonic excitation

The proper choice of an efficient input excitation is always of benefit for a system identification. Types of different excitation forces like periodic, impulse, and random have been widely studied for the community of modal testing (Ewins, 1984). Other types of designed periodic deterministic excitation formats, and corresponding evaluation tools for the identification of a qualitative system, are presented in Pei and Piyawat (2008) and Gloth and Sinapius (2004). It was demonstrated that the proposed frequency and amplitude modulated periodic excitation can effectively unfold the underlying dynamics of a complex nonlinear hysteretic system, especially in detecting the existence of nonlinearities.

All our previous examples of system identification assumed that the force excitation was a monocomponent quasiharmonic signal with a slow-varying (modulated) frequency. However, theoretically the main analytical signal relations between the input and output signals are true for any kind of input excitation. In particular, the input can also be just the composition of two simple harmonics with constant amplitudes and frequencies. From the classic spectral point of view, two harmonics at the input of a linear vibration system will produce two similar frequency output harmonics with scaled amplitudes and shifted phases. The spectral estimation of the dynamic properties of tested systems brings out only two separated constant amplitude and frequency points on the FRF. There is no detailed information about the response function itself or the system modal parameters. The FORCEVIB method, based on the HT, is one method that allows an estimation of the detailed FRF – even in the case of two input harmonics.

To explain the possibility of such a system identification, let us consider a transformation of the analytical signal transmitted through an arbitrary linear dynamic system with a complex FRF $\mathbf{H}(\omega)$. For a linear dynamic system the output is related to the input signal by the well-known spectral equation $\mathbf{X}(\omega) = \mathbf{H}(\omega)\mathbf{Z}(\omega)$, where $\mathbf{X}(\omega)$ is a complex spectrum of the output, $\mathbf{H}(\omega)$ is a complex FRF of the system, and $\mathbf{Z}(\omega)$ is a complex spectrum of the input. The output signal can also be written in analytic signal form by the use of an inverse Fourier transform $X(t) = x(t) + i\tilde{x}(t) = (2\pi)^{-1} \int_0^\infty \mathbf{X}(\omega) e^{i\omega t} d\omega$. Substituting the spectral equation for the input/output relation we will have the same output signal as a function of time (Vainshtein and Vakman, 1983):

$$X(t) = (2\pi)^{-1} \int_0^\infty \mathbf{H}(\omega)\mathbf{Z}(\omega) e^{i\omega t} d\omega.$$

Expanding the FRF $\mathbf{H}(\omega)$ in a Taylor series about the IF $\omega(t)$ we will obtain the following asymptotic series (Vainshtein and Vakman, 1983)

$$X(t) = \mathbf{H}[\omega(t)] A(t)e^{i\varphi(t)} - i\frac{d\mathbf{H}[\omega(t)]}{d\omega} \dot{A}(t)e^{i\varphi(t)} + \ldots, \qquad (10.1)$$

where $\frac{d\mathbf{H}[\omega(t)]}{d\omega}$ is the first derivative of the FRF system (the first derivative of the phase is a group delay) and $A(t)e^{i\varphi(t)}$ is an input signal in the form of an analytic signal. The restricted finite number of terms in the series (10.1) determines the error of this IF method; it depends on the rate of the variation of the envelope and the IF. With a sufficiently slow modulation of the input signal $X(t)$ and slow-varying instantaneous characteristics $A(t)$, $\dot{\varphi}(t)$, we can restrict only the first term of the asymptotic expansion (10.1). The resultant formula will yield a quasistationary relation between the input and output signals of the dynamic system, where the spectral frequency ω is replaced by the IF $\omega(t)$ $X(t) = \mathbf{H}[\omega(t)]Z(t)$. For the quasistationary mode, when

Table 10.1   Extreme and mean values of the envelope and the IF of the biharmonic signal

Signal value	The envelope	The IF
Minimal	$A_{min} = A_1 - A_2$	$\omega_{min} = \omega_1 - \dfrac{\omega_2 - \omega_1}{A_1/A_2 - 1}$
Maximum	$A_{max} = A_1 + A_2$	$\omega_{max} = \omega_1 + \dfrac{\omega_2 - \omega_1}{A_1/A_2 + 1}$
Average	$\overline{A} = \sqrt{A_1^2 + A_2^2}$	$\overline{\omega} = \omega_1$

the frequency modulation can be considered negligibly small, the output change occurs almost simultaneously with the change in frequency of the input signal. In a general case of fast variations in the IF and the amplitude of the input signal, we can take into account the dynamic corrections to the quasistationary solution by using a larger number of terms in the expansion (10.1) describing the transient processes in a dynamic system. In any case, the IF variation of the input signal produces a frequency variation in the output, so the FORCEVIB identification allows us to extract the system modal parameters, including the FRF in the broad range of frequency variation.

Consider an input excitation that consists of two pure harmonics, each with a different amplitude and frequency $z(t) = A_1 \cos \omega_1 t + A_2 \cos \omega_2 t$. The envelope and the IF of such a double-component signal are varying functions of time. Simple known formulas for the extreme and mean values of the envelope and the IF of biharmonics are shown in Table 10.1.

To achieve the largest range of envelope variation, when $0 \leq A \leq A_{max}$, the amplitudes of two harmonics should be close to each other ($A_1 \approx A_2$). A large range variation of the IF can be achieved when $0 \leq \omega \leq \omega_{max}$, so the frequency of the second harmonic should be $\omega_2 \geq A_1 \omega_1 / A_2$.

10.3.1   Linear system model

In the first model $\ddot{x} + 2 \times 0.01\dot{x} + x = \cos 2\pi f_1 t + \cos 2\pi f_2 t$ we use two harmonics to generate a "beating" vibration regime. The frequency of the first harmonic $f_1 = 0.16$ Hz is close to the linear system resonance frequency $1/2\pi \approx 0.159$ Hz; the frequency of the second harmonic, $f_2 = 0.155$ Hz, is chosen to have three full period beatings during the total time of the recorded vibration $T \gg (f_1 - f_2)^{-1}$. Naturally, a linear system does not change a typical "two harmonics" wave nor the spectrum form of the output (displacement) signal relative to the input (excitation) signal (Figure 10.13).

Intermediate results of the HT identification – such as the time segment of the displacement with its envelope and the instantaneous frequencies – are given in (Figure 10.14. The IF of the displacement (Figure 10.14b, dashed line) varies in time around a constant value of the estimated instantaneous undamped natural frequency (bold line).

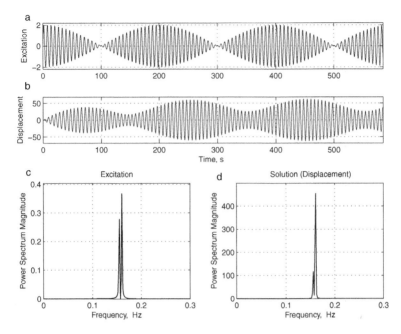

Figure 10.13   Linear vibration system: the excitation (a), the displacement (b), excitation spectrum (c), displacement spectrum (d)

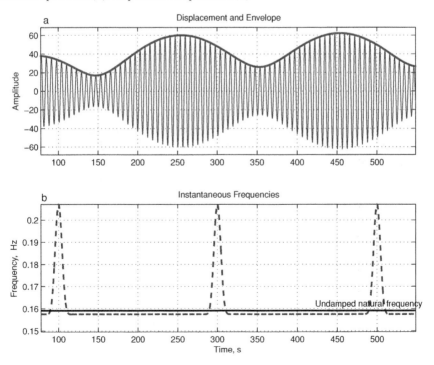

Figure 10.14   Linear vibration system: the displacement and the envelope of the solution (a), the instantaneous frequencies (b)

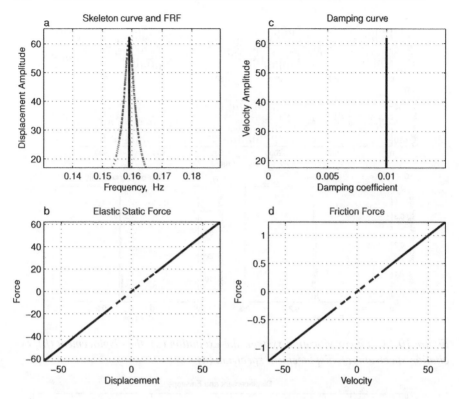

Figure 10.15   The linear system identification: skeleton curve (a, —), FRF (a,...); elastic static force (b); damping curve (c); friction force characteristics (d)

This frequency modulation performance allows us to reconstruct the tested dynamic structure in a wide frequency range. The final results of the HT identification, including the skeleton curve, the FRF, the elastic static force characteristics, the damping curve, and the friction force, are shown in Figure 10.15 by a bold line. In the same figure, the initial characteristics, namely, the linear skeleton line 0.159 Hz, the linear elastic static force $kx = x$, and the linear damping curve $h = 0.01$, are shown by a dashed line. The difference between the initial and estimated linear characteristics is so small (less than 0.1%) that it cannot be distinguished in Figure 10.15.

10.3.2   Nonlinear hardening system

Let us now examine the identification results of a nonlinear system that contains two different types of nonlinearity: nonlinear "quadratic" damping and a nonlinear cubic elastic component inherent to the Duffing equation $\ddot{x} + 2 \times 0.01\dot{x} + 0.2\dot{x}|\dot{x}| + x + 0.4x^3 = \cos 2\pi f_1 t + \cos 2\pi f_2 t$. Again, a forced vibration regime is produced by the same quasiperiodic force input signal. The existence of the nonlinearity can be noticed immediately from the time data ((Figure 10.16a,b), whose input and output

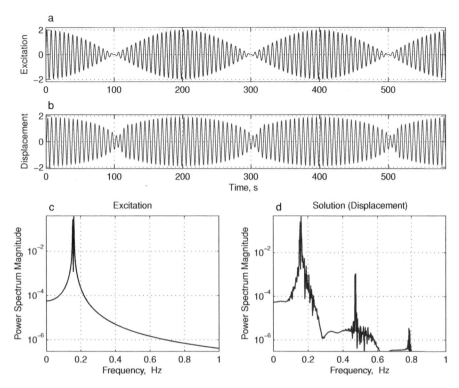

Figure 10.16   Nonlinear hardening system: the excitation (a), the displacement (b), excitation spectrum (c), displacement spectrum (d)

wave shapes differ, and also from the output spectrum (Figure 10.16d), which has high-frequency multiple harmonics. The HT identification results are given in Figures 10.17 and 10.18. A comparison between the estimated (bold line) skeleton curve and the initial average (dashed line) skeleton curve, taken as the first term of (8.9), $f_0(A) = \sqrt{\omega_0^2(A)}/2\pi = \sqrt{\alpha_1 + \frac{3}{4}\alpha_3 A^2}/2\pi = \sqrt{1 + 3A^2}/2\pi$, shows that these skeleton curves are in a good close agreement.

The identified (bold line) and initial precise (dashed line) static force characteristics $kx + \alpha x^3 = 1 + 0.4x^3$ (Figure 10.18b) are very close. The identified (bold line) and initial precise (dashed line) friction force characteristics $0.02\dot{x} + 0.2\dot{x}|\dot{x}|$ are also close to each other (Figure 10.18d).

This model illustrates that the HT identification makes it possible to restore nonlinear characteristics even in the case of a combined nonlinearity in both the elastic and friction parts of the equation of motion under biharmonic excitation. Note that the identified static force characteristics have a small deviation from the initial characteristics to the "linear" direction. In other words, they are slightly less nonlinear than the initial force characteristics, which means that the FREEVIB and FORCEVIB identification procedures restore only the first main term of the nonlinear motion.

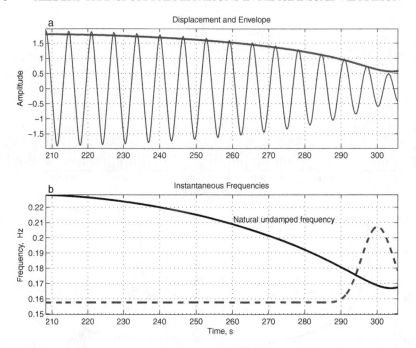

Figure 10.17   Nonlinear hardening system: the displacement and the envelope of the solution (a), the instantaneous frequencies (b)

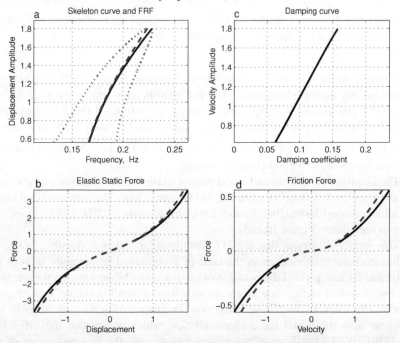

Figure 10.18   The nonlinear hardening system identification: skeleton curve (a, —), FRF (a,...); elastic static force (b); damping curve (c); friction force characteristics (d)

10.3.3   Nonlinear softening system

As an example we refer to the simulation of a vibration system with a nonlinear softening spring and linear friction characteristics $\ddot{x} + 2 \times 8.5\dot{x} + (30 \times 2\pi)^2 x - 1800x^3 = z$. The excitation is a weighted sum of two harmonics, each with its own constant frequency and amplitude value: $z = \cos 2\pi 30t + 0.99 \cos 2\pi 31.5t$ (Figure 10.19a). The simulated forced vibration carried out by the beating excitation is shown in Figure 10.19b together with the envelope function. The alternating IF of the solution, the instantaneous modal frequency of the system, and the phase shift between the input and output are shown in Figure 10.19c.

The obtained backbone has a typical nonlinear form of softening stiffness, and the damping curve is a trivial vertical line (Figure 10.20). The spring and damping force characteristics, estimated after a FORCEVIB identification, practically coincide with the initial ones (Figure 10.21). It was demonstrated that the described identification method based on the HT is able to estimate the modal parameters and the detailed FRF in the range of frequencies – even in the case of two input pure harmonics.

In a simple case, such as the sum of two sinusoidal signals with constant frequencies, only two separate frequency points of the FRF can be estimated by the traditional

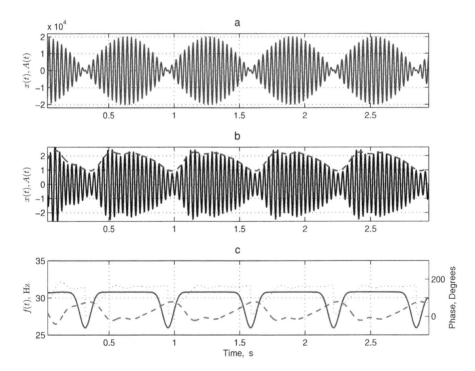

Figure 10.19   The softening system with biharmonics force excitation (a); the displacement solution and the envelope (b); the IF of the swept excitation (—), the instantaneous modal frequency (- -), the phase shift between input and output (. . .)

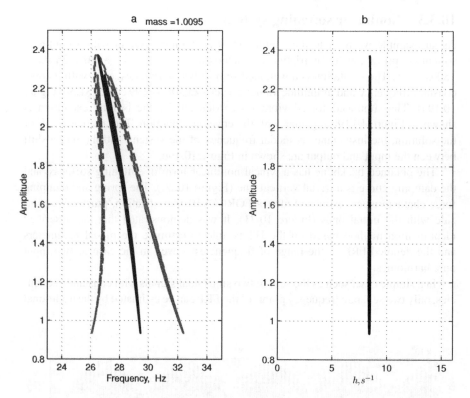

Figure 10.20   The identified Duffing equation under biharmonics force excitation: the backbone (a, —), the FRF (a, - -); the damping curve (b)

Fourier transform method. However, the HT of the same signal contrastingly allows us to estimate the FRF for a wide continuous frequency range.

10.4   Identification of nonlinear time-varying system

The FREEVIB and FORCEVIB methods proposed for identifying instantaneous modal parameters prove to be very simple and effective. They allow the identification of linear time-varying natural frequencies and damping characteristics (Shi and Law, 2007) as well as nonlinear parameters, including their dependencies on the vibration amplitude and frequency. This section concentrates on the dynamic analysis and identification of two groups of dynamic systems: (1) combined forced vibrations of quasiperiodic time-varying linear and nonlinear SDOF systems excited by a harmonic signal; and (2) combined self-excited and forced vibrations of nonlinear SDOF systems excited by a harmonic signal (Feldman, 2005).

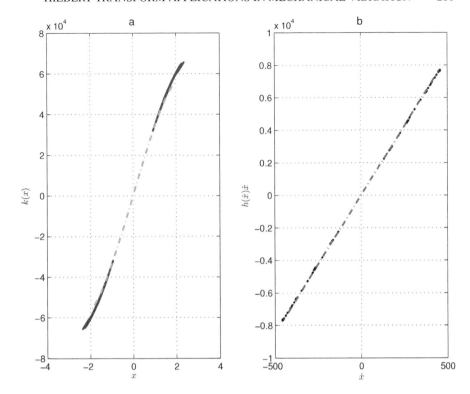

Figure 10.21   The identified Duffing equation under biharmonic force excitation: the spring force (a); the damping force (b)

For the chosen linear and nonlinear models of vibration motion, we use the following analytical expression with the combined different equations:

$$\ddot{x} + h\dot{x} + \delta\dot{x}|\dot{x}| + \mu\dot{x}(\dot{x}^2 - 1) + kx + \alpha x^3 = z(t); \quad k = \omega_0^2(1 + \beta\cos\omega_\beta t);$$
$$z(t) = A_1\cos 2\pi f_1 t + A_2\cos 2\pi f_2 t + A_3\cos 2\pi f_3 t^2 \tag{10.2}$$

Consequently, we will get the vibration motion by forming a combination of the following parameters: h, the linear viscous damping coefficient; δ, the nonlinear "quadratic" friction coefficient; μ, the friction coefficient of the van-der-Pol equation; k, the static elastic force coefficient; ω_0^2, the linear undamped natural frequency square; α, the cubic coefficient of the Duffing equation; β, the amplitude modulation coefficient of the elastic force coefficient; ω_β, the frequency modulation coefficient of the elastic force coefficient; A_i, the amplitude of the excitation; and f_i, the frequency of the excitation. Their corresponding numeric values are given in Table 10.2. The simulations of all differential equations of motion are performed with SIMULINK (MATLAB) with the permanent step value ODE4 (Runge–Kutta) solver.

Table 10.2 Model parameters

| System model | Damping | | | Elasticity | | | | Quasiperiodic | | | | Sweep | |
	h	δ	μ	ω_0^2	α	β	ω_β	A_1	f_1	A_2	f_2	A_3	f_3
1 Modulated elasticity	0.02	0	0	1	0	0.5	0.03	1	0.16	0	0	0	0
2 Modulated elasticity + swept excitation	0.02	0.2	0	1	0	0.5	0.03	0	0	0	0	1	1e^{-5}
3 Parametric excitation	0.02	0	0	1	0	0.5	1	1	0.16	0	0	0	0
4 Van-der-Pol + Duffing	0.02	0	0.1	1	0.4	0	0	1e^{-3}	0.16	0	0	0	0
5 Van-der-Pol + biharmonic excitation	0.02	0	0.1	1	0	0	0	1	0.16	1	0.155	0	0
6 Van-der-Pol + swept excitation	0.02	0	0.1	1	0	0	0	0	0	0	0	0.1	1e^{-5}

The table spans: System parameters (Damping, Elasticity) and Excitation parameters (Quasiperiodic, Sweep).

10.4.1   Model 1. Modulated elasticity

Let us consider a structure with a known unit mass value that describes a slow modulated elastic force $k = \omega_0^2(1 + \beta \cos \omega_\beta t) = 1 + 0.5 \cos 0.03t$ of a trivial dynamics system under an external harmonic excitation (Table 10.2).

The generated vibration has a rather complicated form over time and also in the frequency domain (Figure 10.22). But the HT identification restores this modulation in detail. Thus, Figure 10.23b shows that the identified instantaneous undamped natural frequency of the vibration (bold line) completely coincides with the varying initial elastic force modulation function $f = \sqrt{1 + 0.5 \cos 0.03t}/2\pi$ (dashed line).

The obtained damping characteristics (Figure 10.24, bold line) demonstrate a linear type of friction force, which is a good match for the initial linear type of Model 1 (dashed line) with $\gamma/2 = 0.01$ and $\gamma \dot{x} = 0.02\dot{x}$. This example shows that the HT method makes the identification of a slow modulation of the system parameters possible even in a case of the simplest monoharmonic excitation.

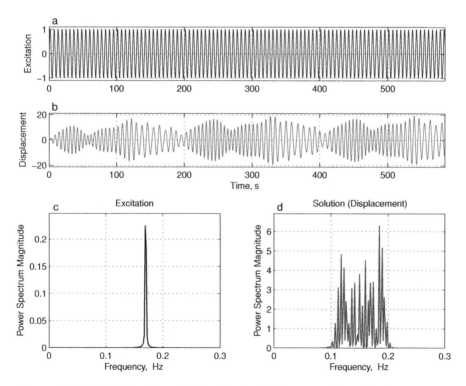

Figure 10.22   Model 1: the excitation (a), the displacement (b), excitation spectrum (c), displacement spectrum (d)

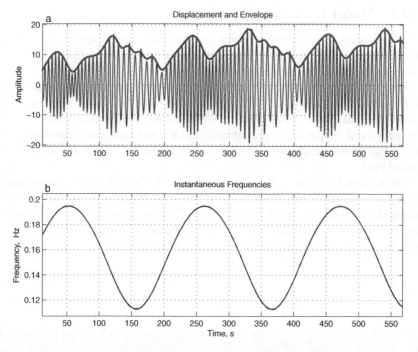

Figure 10.23   Model 1: the displacement and envelope of the solution (a), the IF (b)

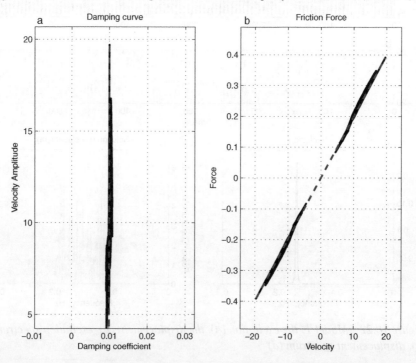

Figure 10.24   Model 1: the identified damping curve (a), the friction force characteristics (b)

10.4.2   Model 2. Modulated elasticity + Quadratic damping + Swept excitation

The next system, as well as the previous one, has a slow modulated elastic force, but in addition to the linear viscosity, it also has a nonlinear quadratic friction member. Also, the tested model has a different external force excitation with a sweeping increasing frequency instead of a single harmonic (Table 10.2). The generated vibration in time, and also in the frequency domain (Figure 10.25), takes a rather complicated form. The results of the HT identification are shown in Figures 10.26 and 10.27.

The identified instantaneous undamped natural frequency (bold line) completely coincides with the initial elastic force modulation $f = \sqrt{1 + 0.5\cos 0.03t}/2\pi$ (Figure 10.26b, dashed line). The obtained nonlinear friction force characteristics (bold line) are in close agreement with the initial nonlinear type of friction force (dashed line) $h\dot{x} + \delta\dot{x}|\dot{x}| = 0.02\dot{x} + 0.2\dot{x}|\dot{x}|$ (Figure 10.27b).

10.4.3   Model 3. Parametric excitation

The static elastic force for this case is modulated by a high modulation frequency equal to the constant natural frequency of the system $\omega_0 = \omega_\beta = 1$ (Table 10.2).

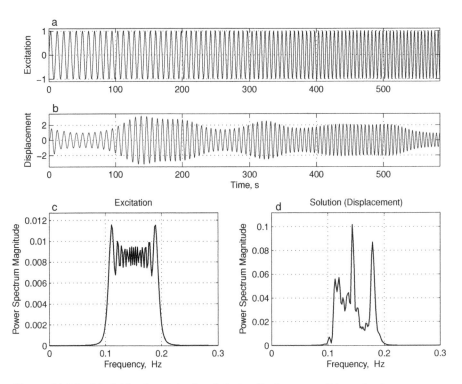

Figure 10.25   Model 2: the excitation (a), the displacement (b), excitation spectrum (c), displacement spectrum (d)

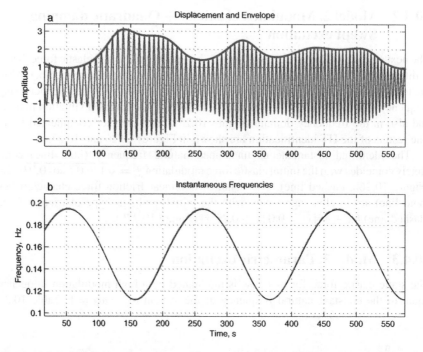

Figure 10.26   Model 2: the displacement and the envelope of the solution (a), the IF (b)

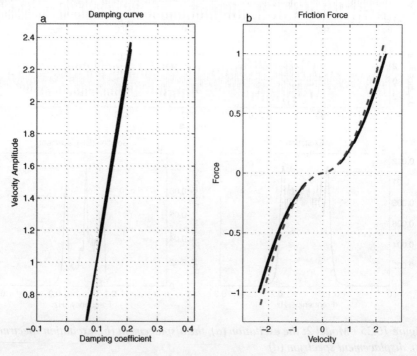

Figure 10.27   Model 2: the identified damping curve (a), the friction force characteristics (b)

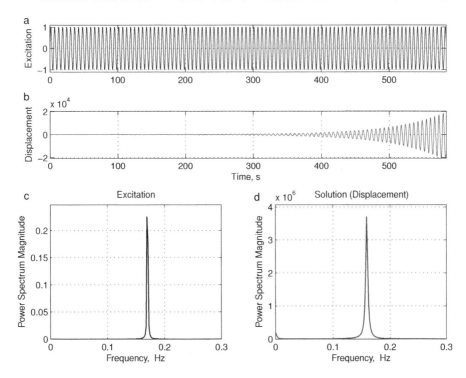

Figure 10.28   Model 3: the excitation (a), the displacement (b), excitation spectrum (c), displacement spectrum (d)

The amplitude of the solution of the system under an external harmonic excitation increases infinitely (Figure 10.28b). This behavior illustrates the parametric instability of the tested system.

In this case, the input harmonic excitation has almost no influence on the observed unstable oscillation. Therefore, we will use FREEVIB method of identification which analyzes only the structure vibration output. The HT identification method restores only the correct skeleton curve and the elastic force characteristics (Figure 10.29a, b). It is clear that, in this case, the HT – instead of the initial linear damping – restores a negative increment and corresponding negative (in the opposite direction) friction characteristics (Figure 10.29c, d). In some special cases, the obtained increment can be used for a quality analysis of the instability growth rate of those unstable vibration solutions.

10.4.4   Model 4. Van-der-Pol + Duffing

The next test combines a nonlinear friction part common to the van-der-Pol oscillator and a nonlinear cubic elastic force part typical to the Duffing equation. Model 4 has a very low level harmonic input excitation (Table 10.2). As expected, it displays a known self-excited regime of nonlinear vibration in time (Figure 10.30b).

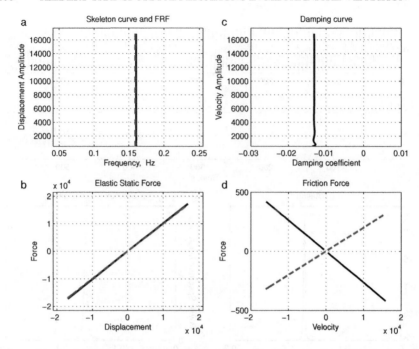

Figure 10.29   Model 3: the skeleton curve (a), the elastic static force (b), the damping curve (c), the friction force characteristics (d)

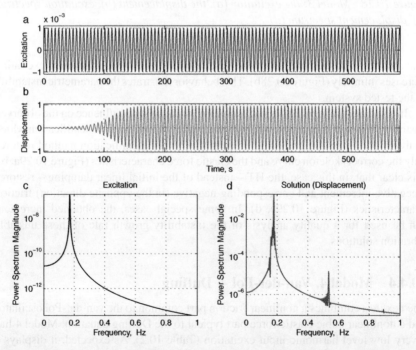

Figure 10.30   Model 4: the excitation (a), the displacement (b), excitation spectrum (c), displacement spectrum (d)

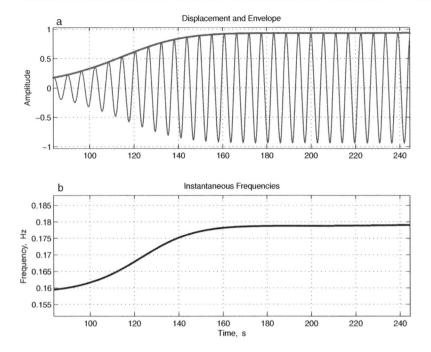

Figure 10.31   Model 4: the displacement and the envelope of the solution (a), the IF (b)

The corresponding spectrum of the self-excited vibration shows high-frequency multiple harmonics, which confirms the existence of the nonlinear elastic part (Figure 10.30d). Again, the observed self-excited oscillation regime does not depend on the input excitation signal.

The HT identification of the FREEVIB method used here (Figure 10.31), restores both the nonlinear friction and nonlinear elastics parts in full detail. Thus, the identified (bold line) skeleton curve and the initial (dashed line) skeleton curve, as the first term of (8.9), almost coincide (Figure 10.32a). Similarly, the identified (bold line) and initial (dashed line) static force characteristics $kx + \alpha x^3 = 1 + 0.4x^3$ also almost coincide (Figure 10.32b). The identified nonlinear friction force characteristics (bold line) are in close agreement with the initial nonlinear type of friction force (dashed line) $\mu\dot{x}(\dot{x}^2 - 1) = 0.1\dot{x}(\dot{x}^2 - 1)$ (Figure 10.32d).

10.4.5   Model 5. Van-der-Pol + Biharmonic excitation

The next system combines a nonlinear friction part that is common to the van-der-Pol oscillator with a high-level quasiperiodic excitation (Table 10.2). The chosen forced vibration regime becomes dominant, and the vibration shape shown in (Figure 10.33a, b is similar in appearance to the forced vibration of the nonlinear system in Figure 10.16a, b. The HT, however, detects the actual properties of the tested system (Figure 10.34). The identified skeleton curve and the elastic static force characteristics

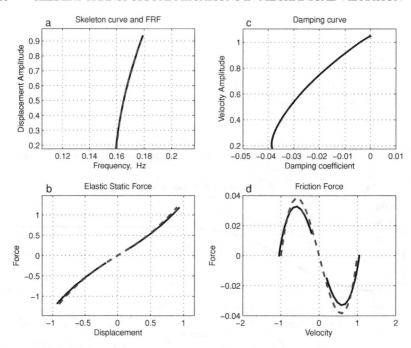

Figure 10.32   Model 4: the skeleton curve (a), the elastic static force (b), the damping curve (c), the friction force characteristics (d)

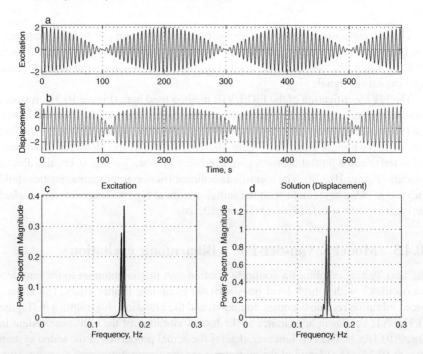

Figure 10.33   Model 5: the excitation (a), the displacement (b), excitation spectrum (c), displacement spectrum (d)

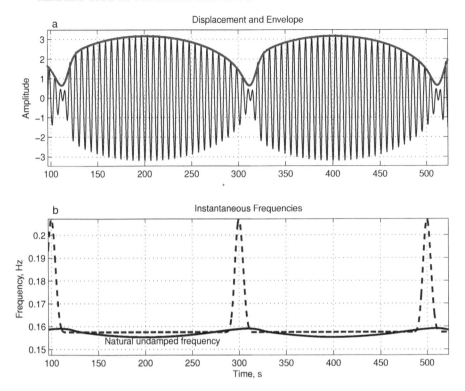

Figure 10.34   Model 5: the displacement and the envelope of the solution (a), the instantaneous frequencies (b)

(bold line) take a trivial linear form corresponding to the initial linear elastics part (dashed line) of the tested system (Figure 10.35a, b). The identified nonlinear friction force characteristics (bold line) are in close agreement with the initial nonlinear type of friction force (dashed line) $\mu\dot{x}(\dot{x}^2 - 1) = 0.1\dot{x}(\dot{x}^2 - 1)$ (Figure 10.35d).

10.4.6   Model 6. Van-der-Pol + Swept excitation

The system under consideration repeats the previous system with a nonlinear friction part that is common to the van-der-Pol oscillator, but now it involves a forced excitation with a sweeping frequency (Table 10.2). The obtained waveform in time, and also the shape of the corresponding spectrum, demonstrate the typical resonance performance of the structure (Figure 10.36). Application of the FORCEVIB identification method is shown in (Figure 10.37). Again the identified skeleton curve and the elastic static force characteristics (bold line) take a trivial linear form that corresponds to the initial linear elastics part (dashed line) of the tested system (Figure 10.38a, b). The identified nonlinear friction force characteristics (bold line) are in close agreement with the initial nonlinear type of friction force (dashed line) $\mu\dot{x}(\dot{x}^2 - 1) = 0.1\dot{x}(\dot{x}^2 - 1)$ (Figure 10.38d).

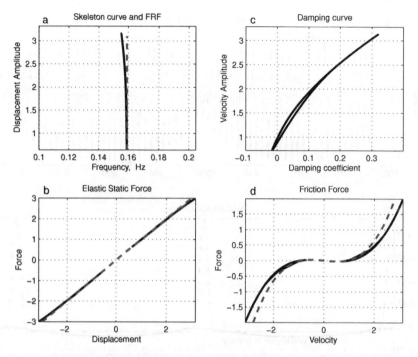

Figure 10.35   Model 5: the skeleton curve (a), the elastic static force (b), the damping curve (c), the friction force characteristics (d)

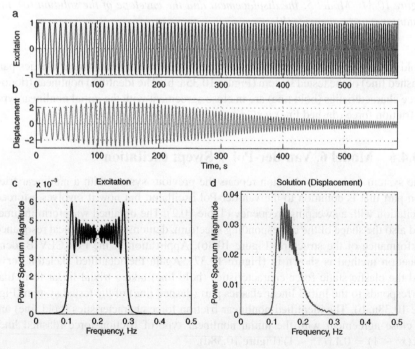

Figure 10.36   Model 6: the excitation (a), the displacement (b), excitation spectrum (c), displacement spectrum (d)

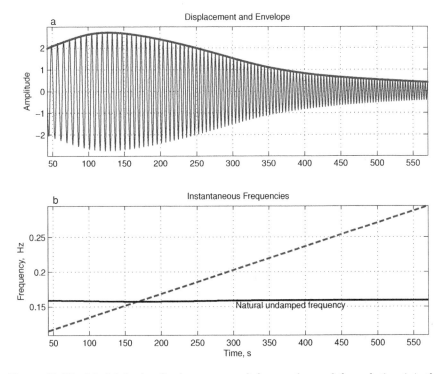

Figure 10.37   Model 6: the displacement and the envelope of the solution (a), the instantaneous frequencies (b)

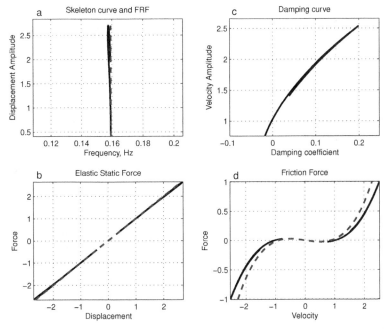

Figure 10.38   Model 6: the skeleton curve (a), the elastic static force (b), the damping curve (c), the friction force characteristics (d)

We can conclude that the used HT identification methods make it possible to reconstruct, in detail and separately, the actual combined nonlinear elastic and friction force of the equation of motion of SDOF systems. The proposed HT technique allows us to identify the time-varying (modulated) parameters of linear and nonlinear systems under different kinds of excitation – including the quasiperiodic input signal – with only two harmonic components. The HT identification allows us to reconstruct the tested dynamic structure in a wide continuous frequency range around the resonance, whereas the traditional Fourier transform only permits us to estimate two discrete frequency points on the FRF. A further continuous real-time estimation of the initial parameters of nonlinear dynamical systems can be used, for example, for measurement, monitoring, and diagnostics purposes. Recent work in the area of the time-domain representations of vibration, such as an HT analysis, shows great promise for the identification of dynamic systems.

10.5   Experimental Identification of nonlinear vibration system

The provided measurements of a free and forced vibration motion and unique signal processing, based on the HT analysis, yield an accurate estimation of nonlinear spring and friction parameters of the vibration model. The obtained natural frequencies and friction parameters are functions (rather than scalars) that describe the system's behavior under different operating conditions. The important type of nonlinearity arises when the restoring force of the spring is not proportional to its deformation. There are several known types of static force characteristics (load–displacement curve) representing different types of nonlinearity in elastic springs: backlash, preloaded (precompressed), impact, and polynomial. A vibrating system, normally described by fixed parameters, can be presented by a piecewise-linear restoring force that may also be considered as an approximation of continuous typical curves (Worden and Tomlinson, 2001). In most nonlinear vibration systems the natural frequency will be decisively dependent upon the vibration amplitude. Therefore typical nonlinearities in springs have the unique form of a skeleton (backbone) curve. The topography of the skeleton curve is essential to assess the properties of the tested vibrating system. The resultant characteristics of nonlinear elastic forces can be reconstructed on the final stage of the HT identification on the basis of the estimated modal parameters.

The nonlinear spring force function, identified from measured vibrations of a SDOF system, corresponds well to the system's initial static elastic force characteristics. In general, even a SDOF nonlinear system could include several elastic and damping elements that can be combined integrally through parallel and/or series connections. In these complicated cases of system identification, there is no unique solution for the decomposition of the resultant force characteristics, and one should use some additional information about the model structure and the combination of its elements.

10.5.1   The structure under test

An experimental vibrating structure was made with the following features: the structure consisted of a mass attached to a heavy base by means of two springs, as shown in Figure 10.39. The mass between the springs could move horizontally about its equilibrium position, while the springs vibrated as fixed-end beams. A test rig included a special pretensioning mechanism, coupled to a base plate. The experimental vibrating system included an external actuator allowing us to apply a variable force excitation and a LVDT sensor to measure the mass motion.

In this work the identification was carried out on the basis of experimentally determined instantaneous characteristics of free and forced vibration response

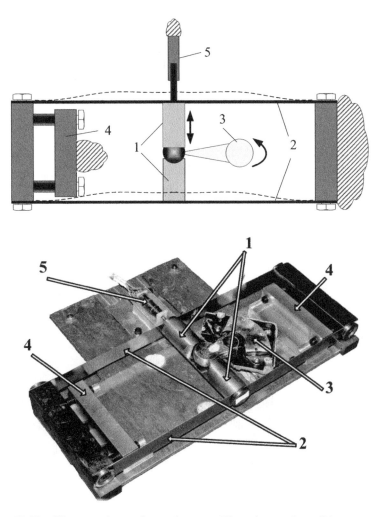

Figure 10.39   The experimental stand: mass (1), ruler springs (2), actuator (3), tension mechanism (4), LVDT sensor (5)

signals measured from the test stand. By applying a nonparametric HT identification technique and an instantaneous signal frequency estimation, one can compute the signal envelope and produce the structural parameters. These steps do not pose serious computational or procedural difficulties and can be performed in a short time for an arbitrary type of nonlinearity inherent in the system.

10.5.2   Free vibration identification

The free vibration displacement signal was produced by abruptly stopping a forced excitation that was exciting the system at resonance (here, 13 Hz). The actuator – under free vibration – repeatedly generated an excitation sequence burst consisting of periodic excitation (at 13 Hz) that stopped for a short period, then started again. Figure 10.40 shows an example of four repeated patterns of the measured displacement and Figure 10.41 shows the forced excitation regime. Naturally, the free vibration decay corresponds only to the decaying part of each pattern that takes place when the actuator is switched off. The experimental investigation shows that the tested structure is represented only by a SDOF system.

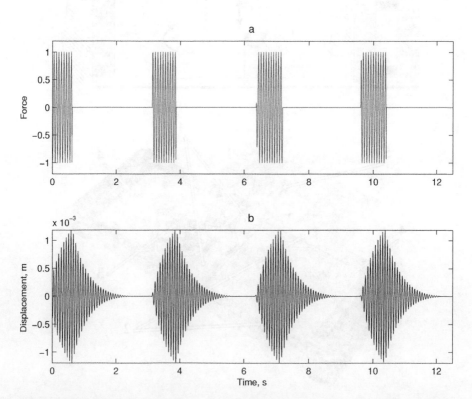

Figure 10.40   The measured time histories: the repeated interrupted force excitation (a), the output free vibration displacement (b)

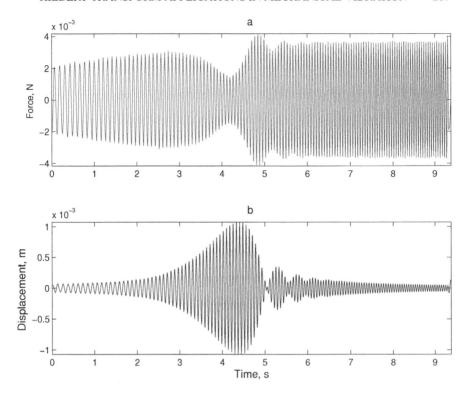

Figure 10.41   The measured time histories: the input sweeping force excitation (a), the output displacement (b)

The backbones of four repeated impulses obtained according to (9.4) are shown concurrently in Figure 10.42 (dash-dot line) along with the corresponding amplitude response according to (9.9). All these backbones almost coincide, which indicates that the natural frequency is an amplitude-dependent function. The skeleton curve tips out of vertical to the right for both small (less than $3 \cdot 10^{-3}$ m) and large amplitudes (more than $5 \cdot 10^{-3}$ m). That means that the tested structure includes two different types of nonlinear stiffness elements – the preloaded amplitudes and the hardening spring. For the large amplitudes and hardening spring, a polynomial curve fit (8.9) of the estimated nonlinear skeleton curve gives the following form $\omega^2(A) = 4\pi^2 12.67^2 + \frac{3}{4} 0.46A^2$ [rad/s]^2. For the small amplitudes from the same skeleton curve a very small preloading force value equal to $F_0/k = 1.2 \cdot 10^{-5}$ [m] was estimated according to (8.2).

The obtained damping curves for four excitation patterns almost coincide, showing little variation in their estimated average damping coefficient $h = 2.5$ [1/s] (Figure 10.44a). Estimation of the damping allows us to construct a curve showing both the amplitude of the damped response for a free vibration regime, and the damping static force characteristics (Figure 10.44b). As can be seen, the damping force characteristics indicate that there exists a combination of linear viscous friction and dry friction.

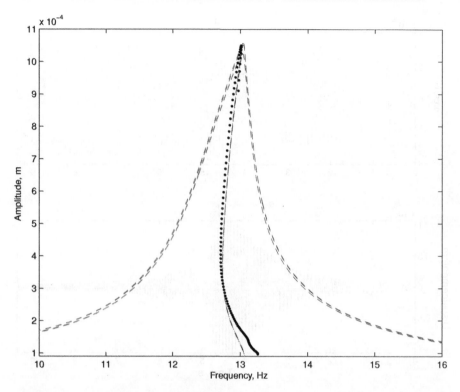

Figure 10.42   The experimental skeleton curves and frequency response: the skeleton curves of free vibrations (---), the frequency response functions (- -), the skeleton curves of forced vibrations (···)

The dry friction force per unit mass is obtained by performing a polynomial curve fitting: it is equal to 0.015 [m/s²].

The resultant identified model per unit mass takes the form of a SDOF vibration system

$$\ddot{x} + 5\dot{x} + 0.015\,\text{sgn}(\dot{x}) + (2\pi\,12.67)^2\,x + 0.46x^3$$
$$+1.2 \cdot 10^{-5}\,(2\pi\,12.67)^2\,\text{sgn}(x) = 0, \qquad (10.3)$$

for the range $0 < A_x < 1.5 \cdot 10^{-3}$ [m], it is clear that the nonlinear spring and nonlinear damping cannot be ignored in both the large and small amplitude range.

The results of the HT identification of a free vibration (10.3) describe, for the main system, the linear and nonlinear properties including the skeleton and damping curves, as well as the relative static stiffness and damping force characteristics per unit mass. These results constitute a basis for model identification, but a free vibration analysis does not restore the absolute system mass and stiffness values. To estimate the absolute values of the system parameters we provide a forced vibration regime of the system.

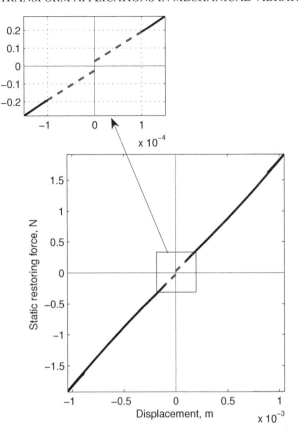

Figure 10.43   The experimental stiffness static force characteristics

10.5.3   Forced vibration identification

During this test, the forced vibrations were produced by the actuator (exciter) with a continuous frequency sweep in the range of 5 to 20 Hz. The force generated by the actuator is shown in Figure 10.41a, and the corresponding forced vibration is shown in Figure 10.41b. The HT identification method FORCEVIB uses these input and output time histories, where the displacement and the force are presented in the time domain.

The obtained results, as well as the results from the free vibration identification, evidently include the same skeleton and damping curves, and also the same static stiffness and damping force characteristics. For example, Figure 10.42 (dotted line) shows the skeleton curve, which almost coincides with the same curve from the free vibration regime. However, the forced vibration identification was able to restore the reduced mass absolute value and the stiffness absolute value. The obtained absolute mass value of all the moving parts of the experimental stand is equal to 0.27 kg. The obtained mass and the natural frequency give the value of the structure reduced average static stiffness value: $k = m\omega_0^2 = 0.27\,(2\pi \cdot 12.7)^2 \approx 1.7 \cdot 10^3 [\text{N/m}]$. The

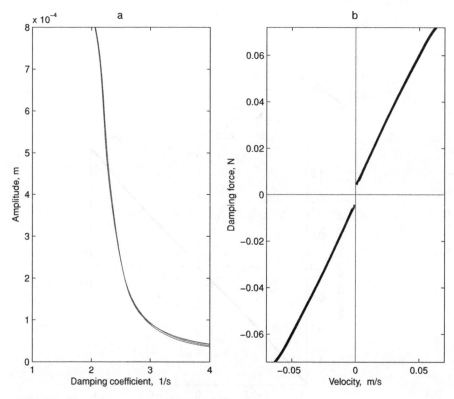

Figure 10.44   The experimental damping curve (a) and the friction static force characteristics (b)

obtained mass value also determines both absolute (static) force characteristics (stiffness and damping, Figures 10.43b and 10.44b) and the final model of forced vibration of the system:

$$0.27 \left[\ddot{x} + 5\dot{x} + 0.015 \, \mathrm{sgn}(\dot{x}) + (2\pi \, 12.67)^2 \, x + 0.46x^3 + {}^{0.02}/_{0.27} \, \mathrm{sgn}(x) \right]$$
$$= F(t), \tag{10.4}$$

$0 < A_x < 1.5 \cdot 10^{-3}$ [m], where 0.27 is an estimated mass value, and $F(t)$ is an external force. The last formula almost repeats (10.3), but now the forced vibration model incorporates an identified mass value. The identified model, having nonlinear elastic and damping forces, describes the motion of the system under different types of input excitation.

In this section we presented the results obtained by the HT nonparametric identification of a real mechanical system consisting of a mass, spring, and damping. The input force and the system displacement response were measured under free vibrations (transient) and also under a harmonic forced excitation. The identification was carried out by means of the HT technique of a signal processing for nonlinear systems. This technique is based on the analysis of the input and output signals of a system: its envelope and its IF in the time domain.

The HT-based technique makes it possible to estimate directly the system's instantaneous dynamic parameters (i.e., natural frequencies, damping characteristics) and their dependence on the vibration amplitude and frequency from the measured time signals of the input and output. The model of the tested structure was created by curve fitting the experimentally obtained force characteristics. The obtained results may be used to verify and validate the model under different conditions, to simulate possible solutions generated by any other input force, and to find a control scheme that provides the desired vibration response. The introduced identification method of a free and force vibration analysis, which determines instantaneous modal parameters, contributes to an efficient and more accurate testing of nonlinear oscillatory systems, avoiding time-consuming measurement and analysis.

10.6   Conclusions

In conclusion, it can be stated that an interesting and promising experimental method for the identification of nonlinearities in the stiffness and damping characteristics of a vibration system has been developed. The method is based on input and output time-domain measurements and on their Hilbert transforms. The method defines the instantaneous modal parameters (backbones, damping dependencies) of a system under a slow or a very fast swept frequency test

These HT identification methods, named FREEVIB and FORCEVIB, enable us to perform a detailed reconstruction and separation of the actual combined nonlinear elastic and friction force of the equation of motion of a SDOF system. The proposed HT technique permits the identification of linear, nonlinear, and modulated parameters under different kinds of excitation, including a quasiperiodic input signal with only two harmonic components. The HT identification allows for a reconstruction of the tested dynamic structure in a wide continuous range around the resonance frequency, whereas the traditional Fourier transform enables the estimation of only two corresponding discrete frequency points on the FRF.

The HT approach is suitable for any vibration system and does not require knowledge of the signal or the parameters of the system. Such a nonparametric identification will determine not only the amplitude and frequency dependencies, but also the initial nonlinear restoring and damping forces.

The HT methods allow the testing time of prototypes to be reduced without diminishing the accuracy of the data. A further continuous real-time estimation of the initial parameters of nonlinear dynamical systems can be used, for example, for measurement, monitoring, and diagnostic purposes.

11

Considering high-order superharmonics. Identification of asymmetric and MDOF systems

During the previous two decades the HT has been applied more and more widely for the study and identification of nonlinear vibration systems in the time domain. The first results showed that the estimated envelope and the IF, as well as the identified modal parameters, are combinations of slow-varying and fast-oscillated functions (Feldman, 1985; Davies and Hammond, 1987). The fast oscillations and modulations observed are not the results of the influence of the HT signal-processing procedure, because the results of the HT identification of linear systems are smooth and have no oscillations. The HT just reflects the nonlinear nature of an observed vibration solution. The fast-oscillating nonlinear distortions of vibrations are caused by deviations in the linear relationship between the input and output of a dynamical system.

To achieve traditional smooth skeleton curves and static force characteristics, a lowpass filtering was proposed for the approximation of nonlinear system identification results (Feldman, 1994a, 1994b). This lowpass filtering or averaging approach simply eliminates (or ignores) the obtained fast oscillations. This had already been observed in the FREEVIB and FORCEVIB identification methods that operate only with the primary vibration solution of a SDOF system. However, because of the non-linearity in addition to the primary vibration solution, there could be a wide variety of secondary nonlinear high-order *superharmonics* (*ultraharmonics*) with frequencies that differ from the primary resonance response. In reality a vibration solution is generally rich in harmonic content with the major component having a primary natural frequency and a number of minor components having higher multiple frequencies.

Hilbert Transform Applications in Mechanical Vibration, First Edition. Michael Feldman.
© 2011 John Wiley & Sons, Ltd. Published 2011 by John Wiley & Sons, Ltd.

Considering the time domain, the rich harmonic content means the intrawave amplitude and phase modulation of the instantaneous parameters, such as the solution amplitude and frequency.

We can provide a new analysis and a precise identification of nonlinear vibration structures, based on the HT signal decomposition, by joint consideration of the primary and high harmonics solution. The analysis focuses on nonlinearities and on an identification of precise modal parameters (natural frequencies, damping, and static force characteristics) for the free and forced vibration of a SDOF system. Recent achievements in nonstationary signal decomposition (Huang, Shen, and Long, 1999; Feldman, 2006) open the way to extract the fast oscillations and examine them for a more precise nonlinear system identification. Therefore the new analysis will be based on two other HT methods:

- the FREEVIB and FORCEVIB methods for extracting the instantaneous modal parameters;

- the HVD method that splits a nonstationary wideband oscillating signal into separate components.

11.1   Description of the precise method scheme

The HT identification of vibration systems in the time domain is a typical dynamics inverse problem. Suppose the mass of the investigated SDOF dynamics system with unknown restoring and damping forces moves under (or without) an excitation force. By observation (experiment), we acquire knowledge of the position, velocity and acceleration of the object, as well as the excitation at several known instants of time. The inverse problem can be formulated as a question: Can initial restoring and damping forces be precisely determined? In the case of a free vibration we have only an output signal – the vibration of the body; in the case of a forced vibration we deal also with an input excitation.

For linear class systems, such an identification problem has classically known solutions. However, for nonlinear systems, the inverse problem is more complicated, first, because of the intricate nonlinear relationship between the input and the output. In addition, nonlinear systems produce nonlinear harmonic and intermodulation distortions in their output. For example, a harmonic distortion occurs when a system, whose input is fed with a pure sine-wave signal of frequency f, produces as its output a vibration of frequency f, as well as a set of higher harmonics whose frequencies $(2f, 3f, \ldots, nf)$ are harmonically related to the input frequency (Lang and Billings, 2005).

Actually, the real solution of nonlinear systems contains a main (fundamental, primary, principal) sine-wave solution of frequency f along with an infinite number of multiple high-frequency superharmonics nf. Thus a response of the weakly nonlinear system $x(t)$ can be expressed as the sum of the first $x_1(t)$, second $x_2(t)$, and other high-order oscillation functions: $x(t) = \sum_{l=1}^{\infty} x_l(t)$. The FREEVIB and FORCE-VIB identification methods were based only on the extraction of a single principal component of the solution. The performance of other high-frequency superharmonics was ignored or just expressed as a total harmonic distortion coefficient at a specified output level. Considering only the principal component and its corresponding averaged natural frequency, we will get an approximate result correct to the polynomial constant of the stiffness force coefficients (Section 8.3.1). For example, in the case of a pure backlash stiffness such bias difference between the real and the estimated backlash force value will be equal to 25%.

In other cases the initial static force characteristics and those identified by FREE-VIB and FORCEVIB are closer to each other; nevertheless, the identified static force characteristics have a small "natural linearization" deviation from the initial characteristics to the "linear" direction (Figures 10.18 and 10.27). In other words, they are slightly less nonlinear than the initial force characteristics. This means that the HT identification based on the averaging (filtering) procedure restores only the first main term of the motion. This approximation can be recommended mostly to identify nonlinear systems in the case of noisy experimental conditions.

By using modern signal decomposition methods (Huang *et al.*, 1998; Feldman, 2006) we can divide the real multicomponent motion into a number of several separated superharmonics and identify the initial system partially for every component. However, this approach (Feldman, 2006) is suitable only for a free vibration motion that can be decomposed into a principal (fundamental) and other high-order harmonics. Actually, a decomposed multicomponent forced vibration motion cannot be correlated to a single pure sine-wave excitation. Therefore the identification of nonlinear systems requires the development of a more universal method that is suitable for both free and forced vibrations.

11.2   Identification of the instantaneous modal parameters

Typically, a nonlinear equation describing a vibration motion in a SDOF system has a fixed structure. This structure classically includes three independent members: a restoring force (stiffness, spring) $k(x)$ as a nonlinear function of displacement (position), a damping force (friction) $h(\dot{x})\dot{x}$ as a nonlinear function of velocity (the first derivative of the position with respect to time), and an inertial force proportional to acceleration $\ddot{x}$ (the second derivative of the position with respect to time). The second-order vibration system $\ddot{x} + 2h(\dot{x})\dot{x} + k(x) = 0$, having nonlinear elastic and damping force characteristics, can be transformed into the complex equation (9.3):

$$\ddot{X} + 2h(t)\dot{X} + \omega_x^2(t)X = 0, \tag{11.1}$$

where $\omega_x(t)$ is a fast-varying instantaneous modal frequency, $h(t)$ is a fast-varying instantaneous modal damping coefficient, and $X = x(t) + i\tilde{x}(t)$ is the solution. Here the instantaneous modal parameters are fast-varying functions of time. Their direct extraction and plotting demonstrates an unusually fast oscillation (modulation) form.

To identify a nonlinear SDOF system under free vibration, we can measure all the output kinematic parameters – displacement, velocity, and acceleration – as real functions at several time instants. Therefore, at every point in time, the real equation of motion has two unknown parameters: instant stiffness and instant damping.

By applying the HT and introducing an analytic signal with a magnitude and phase representation, we obtain the second equation of motion written for an imaginary HT projection. Thus, instead of only one equation of motion we now have two with the same unknown modal parameters. So we can solve the system directly and estimate the unknown instant stiffness and instant damping for every point of time. Once the stiffness and damping coefficients are known, the spring and damper static force characteristics can be obtained trivially as a multiplication of the corresponding coefficient and motion envelope.

At the first stage of the precise identification technique, the signal envelope, along with the instantaneous natural frequency and the instantaneous damping coefficient, are extracted from the vibration and excitation signals – according to HT signal-processing FREEVIB or FORCEVIB methods but without lowpass filtering. At the next stage, the fast-varying modal parameters are decomposed into separate synchronous components that are combined congruently to produce smooth modal parameters related to the initial static force characteristics. At the final stage, the precise nonlinear static elastic and damping force characteristics are constructed as a multiplication of two corresponding congruent envelopes (for example, the displacement and the elastic force) according to the technique to be described in Section 11.4. It is convenient to represent the final result of the HT identification in a standard form that includes skeleton curves and the initial static force characteristics of the nonlinear vibration system.

11.3   Congruent modal parameters

The measured and estimated instantaneous functions – namely displacement, velocity, natural frequency, and damping – generally turn into fast-varying nonstationary functions due to existing high-order superharmonics. To reconstruct the initial nonlinear force characteristics precisely, we deduce the form of each oscillating function using a suggested term called a congruent envelope (Section 5.6).

11.3.1   Congruent envelope of the displacement

It is clear that precise values of the elastic force function match maximum points of the primary displacement solution $X(t)$. Moreover, the corresponding value of velocity at the maximum displacement point is close to zero, so its contribution to the varying natural frequency is negligibly small. The maximum displacement points correspond to the congruent envelope that is the EOE function of the solution. Such a smooth EOE is a phase congruent function of the primary solution $A_{EOE}(t) = \sum_{l=1}^{N} A_l(t) \cos \phi_l(t)$, where $A_l(t)$ is the envelope of the solution l order harmonic, and $\phi_l(t)$ is the phase angle between the primary and the l order harmonic. For example, if all high-order harmonics are congruently in phase with the primary solution, the EOE of the fast-varying solution is equal to the sum of the harmonic envelopes. It is clear that an estimation of the congruent envelope $A_{EOE}(t)$ requires decomposing the system solution into a sum of high order harmonics. The following algebraic sum of all high harmonic envelopes will produce a congruent envelope.

11.3.2 Congruent modal frequency

In nonlinear vibration systems the obtained instantaneous modal frequency $\omega_x(t)$ also varies fast in time. So the instantaneous natural frequency in its turn can also be decomposed into a sum of high-order synchronous components $\omega_x(t) = \sum_{l=1}^{N} \omega_{xl}(t)$. The main result of such a decomposition is the *congruent modal frequency*

$$\omega_{x\text{EOE}}(t) = \sum_{l=1}^{N} a_{\omega l}(t) \cos \phi_{\omega l}(t) \tag{11.2}$$

estimated as an algebraic sum of the envelopes of all considered high harmonics $a_{\omega l}(t)$. Here $a_{\omega l}$ is an envelope of the l-order harmonic, and $\phi_{\omega l}(t)$ is a phase angle between the primary and the l-order harmonic of the instantaneous modal frequency. The congruent modal frequency as a smooth envelope of an instantaneous modal frequency theoretically defines the exact natural frequency of a nonlinear system for every specified level of the solution. In practice, the accuracy of the last expression depends on the total number N of considered synchronous vibration components. The precise elastic force function is defined at the point of maximum displacement as the multiplication of the congruent displacement envelope and the congruent modal frequency squared.

11.3.3 Congruent modal damping

The instantaneous modal damping $h(t)$ can be decomposed into the sum of high-order synchronous components $h(t) = \sum_{l=1}^{N} h_l(t)$. Thereafter the *congruent modal damping* is estimated as the algebraic sum of the envelopes of all considered synchronous high harmonics $a_{hl}(t)$:

$$h_{\text{EOE}}(t) = \sum_{l=1}^{N} a_{hl}(t) \cos \phi_{hl}(t) \tag{11.3}$$

Here a_{hl} is the envelope of the l-order harmonic, and $\phi_{\omega l}(t)$ is the phase angle between the primary and the l-order harmonic of the instantaneous modal damping. The congruent modal damping as a smooth function theoretically defines the exact damping coefficient of the nonlinear system for every specified level of the solution. Respectively, the multiplication of the congruent modal damping and the congruent velocity envelope at the point of maximum velocity gives the exact value of the friction static force.

11.3.4   Congruent envelope of the velocity

Around every velocity peak point the corresponding value of displacement is close to zero, so its contribution to the varying damping coefficient is negligibly small. The *congruent envelope of the velocity* is a smooth EOE of the solution velocity: $A_{\dot{x}EOE}(t) = \sum_{l=1}^{N} A_{\dot{x}l}(t)\cos\phi_{\dot{x}l}(t)$, where $A_{\dot{x}l}(t)$ is the envelope of the l-order velocity harmonic, and $\phi_{\dot{x}l}(t)$ is the phase angle between the primary and the l-order velocity harmonic.

The EOE has a traditional mathematical meaning as a function (curve) that is tangent to the oscillating envelope. As mentioned, the EOE is a rather smooth function in time; nevertheless, it contains intensities of the primary harmonic and all intrinsic high harmonics existing in the corresponding fast-varying nonstationary function. Calculation of the EOE is based on two cascading operations: a nonstationary signal decomposition and a congruent summation of envelopes and instantaneous functions of the decomposed synchronous components.

11.4   Congruent nonlinear elastic and damping forces

According to (9.10) spring force characteristics $k(x)$ for the identification are defined as the multiplication of two phasors: $K = \omega_x^2 X$, where ω_x^2 is the varying nonlinear natural frequency, X is the displacement of the vibration in a signal analytic form. The precise elastic force function is defined at the point of maximum displacement as the multiplication of the congruent displacement envelope and the congruent modal frequency squared. For a complex product the magnitudes are multiplied and the angles are added together, so the following expression returns the initial smooth static force characteristics:

$$k(x) = \begin{cases} \omega_{xEOE}^2 A_{EOE}, & x > 0 \\ -\omega_{xEOE}^2 A_{EOE}, & x < 0 \end{cases}, \tag{11.4}$$

where ω_{xEOE}^2 is the congruent modal frequency squared, and A_{EOE} is the congruent envelope of the displacement. The accuracy of the last expression depends on the total number of synchronous high harmonics (signal components) considered, and – what is significant – it defines theoretically the exact solution for the identified nonlinear static force characteristics.

By analogy, one can get corresponding expressions for an identified nonlinear damping force in the case of a nonconservative system. Respectively, a multiplication of the congruent modal damping and the congruent velocity envelope at the point of maximum velocity gives the exact value of the friction static force. When the instantaneous damping $h(t)$ is decomposed into the sum of synchronous high-order components, and the congruent modal damping $h_{EOE}(t)$ is also estimated as an algebraic sum of envelopes of all synchronous high harmonics $a_{hl}(t)$, we can get the initial smooth damping force precisely:

$$h(\dot{x})\dot{x} = \begin{cases} h_{EOE} A_{\dot{x}EOE}, & \dot{x} > 0 \\ -h_{EOE} A_{\dot{x}EOE}, & \dot{x} < 0 \end{cases}, \tag{11.5}$$

where h_{EOE} is the congruent modal damping, and $A_{\dot{x}EOE}$ is the congruent envelope of the velocity.

In a general case, a vibration system can have both types of nonlinearities – spring and damping – acting simultaneously. In the case of a combined nonlinear spring and nonlinear damping, the instantaneous natural frequency and the instantaneous damping are combinations of fast-varying cross-components. The energy of the vibration system at each moment is partly kinetic and partly potential, there are moments in time when all the energy is stored mainly as strain energy of the elastic deformation and the fictitious damping is equal to zero. These points correspond to the elastic force and displacement maxima. In other words, phase angles between the main and high-order harmonics naturally split the combination of the high harmonics into two different fast-varying members depending on elasticity or damping. The first member forms a cosine phase projection of the displacement by considering only the elastic force harmonics; the second member independently forms a cosine phase projection of the velocity by considering only the damping harmonics. Therefore, expressions (11.4) and (11.5) are also applicable in the case of a combination of nonlinearities and are common for both types of free and forced vibration. It allows a precise identification of the nonlinear vibration system with a high-order harmonic consideration. In the distinctive view of Hammond and Braun (1986), the congruent modal parameters and the congruent nonlinear elastic and damping forces collectively show not only the geometrical correct wave profile, but also the physically meaningful representation of the full nonlinear system.

11.5   Examples of precise free vibration identification

At the first stage of the proposed identification technique, the envelope $A(t)$ and the IF $\omega(t)$ are extracted from the free vibration using the HT signal processing. Then, the instantaneous undamped natural frequency and the instantaneous damping coefficient of the tested system are estimated according to the formulas (see Section 9.2): $\omega_0^2(t) = \omega^2 - \frac{\ddot{A}}{A} + \frac{2\dot{A}^2}{A^2} + \frac{\dot{A}\dot{\omega}}{A\omega}$; $h_0(t) = -\frac{\dot{A}}{A} - \frac{\dot{\omega}}{2\omega}$, where $A(t)$ and $\omega(t)$ are the envelope and the IF of the vibration.

According to the FREEVIB identification method, a lowpass filtering is applied, thus we obtain only the first (the principal) term of a vibration motion. In this way the averaging restores the approximately correct initial nonlinear forces by using only the first time-varying term of motion. Abandoning the lowpass filtering and considering the secondary and other high-synchronous components of the motion according to (11.2) and (11.3), we will identify the initial nonlinear spring and damping force characteristics precisely.

11.5.1   Nonlinear spring identification

As an example of a nonlinear elastic force, we refer to the classic Duffing equation with a hardening spring and a linear damping characteristic $\ddot{x} + 0.05\dot{x} + x + 0.01x^3 = 0$; $x_0 = 10$, $\dot{x}_0 = 0$. A simulation of a free vibration signal was performed by using an initial displacement, as shown in Figure 11.1. In the same figure, a part of the displacement envelope is shown separately with a zoom to emphasize its oscillating behavior. For example, in the presence of a cubic nonlinearity, the speed

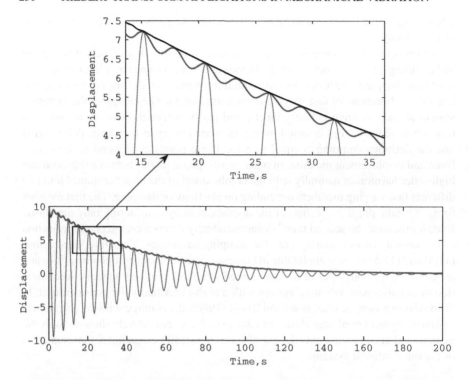

Figure 11.1   The nonlinear spring free vibration: the displacement (—), the envelope (—), the congruent envelope of the displacement (—) (Feldman, ©2009 by John Wiley & Sons, Ltd.)

of oscillations of the envelope and the IF is twice that of the main vibration frequency. It is clear that, because of the nonlinearity, all instantaneous characteristics, including the displacement envelope, the velocity envelope, the instantaneous natural frequency, and the instantaneous damping are fast time oscillating functions.

By applying the HVD method separately to each of the mentioned oscillating functions in time, we will obtain four different decompositions. Each decomposition includes a number of separated high superharmonics of the multicomponent motion. As an illustration, Figure 11.2 shows the first two high superharmonics of the instantaneous natural frequency of a free vibration of the Duffing equation. Each decomposed synchronous component is shown separately in Figure 11.2a as a time history function. A cubic spring nonlinearity occurs essentially in the large amplitude range; therefore, decomposed synchronous components have large envelopes in the range of large vibration levels. Figure 11.2b presents the same IF and envelope of each component in a corresponding 3D plot, where their nonstationary timevarying behavior can be distinctly observed.

Application of the HT identification formulas (11.4) and (11.5) to every synchronous component yields a corresponding part of the nonlinear restoring and damping static force characteristics associated with the sum of the components. The results of the HT identification are shown in Figure 11.3. The fast-varying

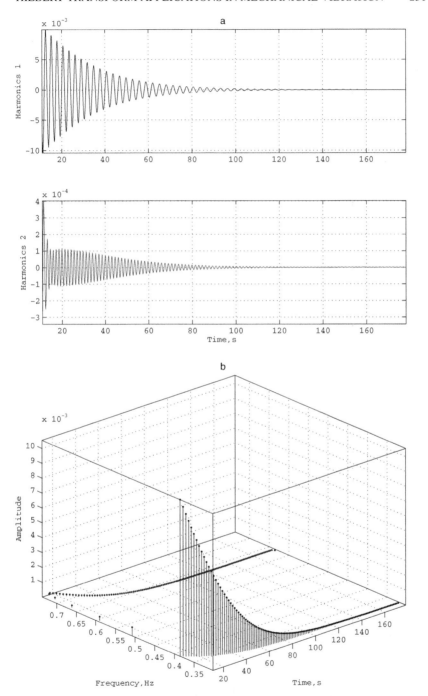

Figure 11.2   The high harmonics of the instantaneous natural frequency of the nonlinear spring free vibration. The time history of two first harmonics (a); the Hilbert spectrum (b) (Feldman, ©2011 by Elsevier)

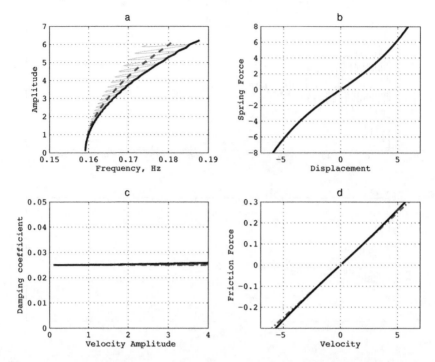

Figure 11.3   The identified parameters of the nonlinear spring free vibration; the skeleton curve (a): instantaneous (—), averaged (- -), identified with high harmonics (—); the spring static force characteristics (b): initial (⋯), identified with high harmonics (—); the damping curve (c): initial (⋯), averaged (- -), identified with high harmonics (—); the friction static force characteristics (d): initial (⋯), identified with high harmonics (—) (Feldman, ©2009 by John Wiley & Sons, Ltd.)

instantaneous natural frequency (before decomposition) is plotted against the envelope as a fast-changing spiral (Figure 11.3a, thin line). The same figure also includes a lowpass filtered skeleton curve (dashed line) that only approximately restores the nonlinear forces correct to the first term of motion. The final precise congruent skeleton curve, which also considers two high superharmonics, is shown in Figure 11.3a (bold line). The resultant identified data (bold line) completely coincides with the initial (dashed line) spring static force characteristics $k(x) = 1 + 0.01x^3$ (Figure 11.3b). Fitting the least-squares data of the static force characteristics to a cubic polynomial model returns a nonlinear coefficient equal to 0.009, which differs by less than 1% from the initial nonlinear coefficient. The resultant identified damping force characteristics (Figure 11.3d, bold line) completely coincide with the initial trivial straight (dashed line) friction force characteristics.

11.5.2   Nonlinear damping identification

The next simulation example, devoted to nonlinear damping, considers the dry friction equation $\ddot{x} + 0.07\mathrm{sgn}(\dot{x}) + x = 0$; $x_0 = 10$, $\dot{x}_0 = 0$. A corresponding free vibration

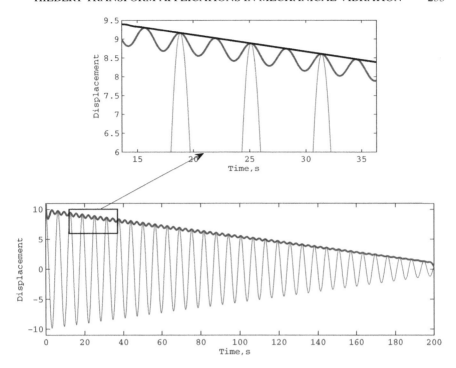

Figure 11.4 Nonlinear friction free vibration: the displacement (—), the envelope (—), the congruent envelope of the displacement (—) (Feldman, ©2011 by Elsevier)

signal with its oscillating envelope is shown in Figure 11.4. The real solution of the nonlinear dry friction system contains principal harmonic and multiple high-frequency superharmonics, so all the associated instantaneous functions are oscillating functions.

Applying the HVD method separately to each of the instantaneous oscillating functions, we obtain the four following decompositions: a displacement envelope, a velocity envelope, a natural frequency, and a damping decomposition. Again, each decomposition includes a number of separated high superharmonics of the multicomponent motion. As an illustration, Figure 11.5 shows the first four high superharmonics of instantaneous damping of the free vibration of the dry friction equation.

These four decomposed components are shown separately in Figure 11.5a, as time history functions. Dry friction nonlinearity occurs essentially in the low-amplitude range; therefore, the decomposed components have large envelopes in the range of small vibration levels. Figure 11.5b presents the IF and the envelope of each of the components in the corresponding 3D plot.

The results of the HT identification, according to formulas (11.4) and (11.5), are shown in Figure 11.6. The instantaneous natural frequency (before decomposition) is plotted against the envelope as a fast-changing spiral (Figure 11.6a, thin line). The same figure includes the final precise congruent skeleton curve, which aggregates the four high superharmonics, shown in Figure 11.6a (bold line). Since the dry friction

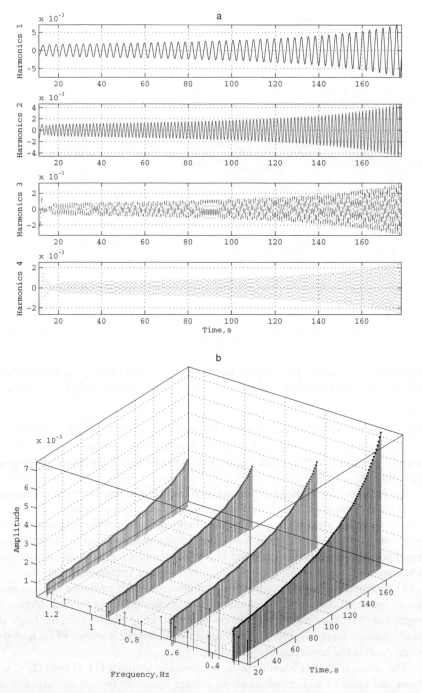

Figure 11.5 *The high harmonics of the instantaneous damping curve of the nonlin-ear friction free vibration: The time history of four first harmonics (a); the Hilbert spectrum (b) (Feldman, ©2011 by Elsevier)*

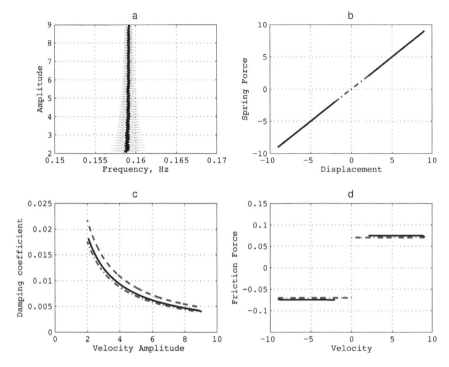

Figure 11.6   The identified parameters of the nonlinear friction free vibration: the skeleton curve (a): instantaneous (—), averaged (- -), identified with high harmonics (—); the spring static force characteristics (b): initial (···), identified with high harmonics (—); the damping curve (c): initial (···), averaged (- - -), identified with high harmonics (—); the friction static force characteristics (d): initial (···), identified with high harmonics (—) (Feldman, ©2011 by Elsevier)

model includes a pure linear spring force, the congruent skeleton curve is a trivial vertical line.

The resultant identified data (bold line) is very close to the initial friction force characteristics $h(\dot{x})\dot{x} = 0.07\text{sgn}(\dot{x})$ (Figure 11.6b, dashed line). A least squares fitting data of the identified friction force characteristics to a horizontal polynomial model returns a nonlinear coefficient of 0.078; this differs by about 1% from the initial coefficient of nonlinear friction.

11.5.3   Combined nonlinear spring and damping identification

As an illustration, consider a nonlinear system with a backlash and a nonlinear turbulent square friction operating in combination:

$$\ddot{x} + 0.07\,|\dot{x}|\,\dot{x} + k(x) = 0; \; k(x) = \begin{cases} x - 0.1\text{sgn}(x - 0.1), \, |x| > 0.1 \\ 0 \qquad\qquad\qquad |x| \le 0.1 \end{cases};$$

$$x_0 = 3, \; \dot{x}_0 = 0.$$

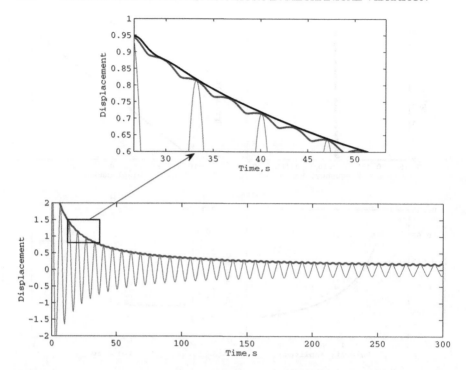

*Figure 11.7   Combined nonlinear spring and damping free vibration: the displace-
ment (—), the envelope (—), the congruent envelope of the displacement (—)*

A corresponding free vibration signal with an oscillating envelope is shown in Figure
11.7. It is well known that the solution of weak nonlinear systems, in addition to
having a principal dominant component, has other small multiple components. All
the obtained instantaneous modal parameters are also oscillating functions.

Applying the HVD method and considering only the first two high superharmonics, we are able to estimate the resultant precise nonlinear restoring and damping force
characteristics of the tested system (Figure 11.8) – according to (11.4) and (11.5).
The fast-varying instantaneous natural frequency (before decomposition), plotted as a
fast-changing spiral with a thin line, forms a kind of solid background in Figure 11.8a.
The lowpass filtered skeleton curve (Figure 11.8a, dashed line) only approximately
restores the nonlinear forces correct to the first term of motion. The final precise
congruent skeleton curve, which takes two high superharmonics into consideration,
is shown in Figure 11.8a by a bold line.

The resultant identified backlash force characteristic (Figure 11.8b, bold line) is
in close agreement with the initial spring static force characteristic $k(x)$ and with a
gap value equal to 0.1 (Figure 11.8b, dashed line). The resultant identified friction
force characteristic (Figure 11.8d, bold line) is very close to the initial quadratic
parabola $h(\dot{x})\dot{x} = 0.07\,|\dot{x}|\,\dot{x}$ (Figure 11.8d, dashed line).

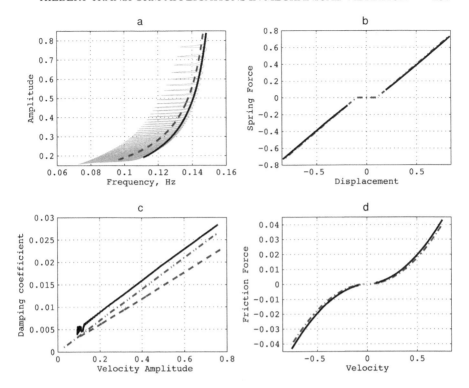

Figure 11.8 The identified parameters of combined nonlinear spring and damping vibrations: the skeleton curve (a): instantaneous (—), averaged (- -), identified with high harmonics (—); the spring static force characteristics (b): initial (···), identified with high harmonics (—); the damping curve (c): initial (···), averaged (- -), identified with high harmonics (—); the friction force characteristics (d): initial (···), identified with high harmonics (—)

11.6 Forced vibration identification considering high-order superharmonics

The differential equation of a weakly nonlinear system under forced excitation $P(t)$ (see Section 9.3) is defined in an analytic signal form: $\ddot{X} + h_0(t)\dot{X} + \omega_0^2(t)X = P(t)/m$, where $h_0(t)$ is the instantaneous damping, $\omega_0(t)$ is the instantaneous natural frequency, $P(t)$ is the forced excitation, and m is the mass of the system. At the first stage of the proposed identification technique, the envelope of the displacement $A(t)$ and the IF of the displacement $\omega(t)$ are extracted from the vibration using HT signal processing. Then, the instantaneous undamped natural frequency and the instantaneous damping of the tested system are estimated according to formulas (9.7):

$$\omega_0^2(t) = \omega^2 + \frac{\alpha}{m} - \frac{\beta\dot{A}}{A\omega m} - \frac{\ddot{A}}{A} + \frac{2\dot{A}^2}{A^2} + \frac{\dot{A}\dot{\omega}}{A\omega}; \quad h_0(t) = \frac{\beta}{2\omega m} - \frac{\dot{A}}{A} - \frac{\dot{\omega}}{2\omega}, \text{ where } \alpha \text{ and } \beta$$

are the following relations between the force input and the displacement output: $\frac{P(t)}{Y(t)} = \alpha + j\beta = \frac{px + \tilde{p}\tilde{x}}{x^2 + \tilde{x}^2} + j\frac{\tilde{p}x - p\tilde{x}}{x^2 + \tilde{x}^2}$, where x is the displacement, $\tilde{x}$ is its Hilbert projection, p is the force, and $\tilde{p}$ is its Hilbert projection.

These expressions for the instantaneous natural frequency and the instantaneous damping consist of two different parts: the forced, or steady-state, part depending on α, β and the free vibration, or transient part. The sum of both parts forms the solution of the general system. By decreasing the rate of external excitation and increasing the data acquisition time, we can reduce the transient part of the motion. The considered forced vibration solution – due to the system's nonlinear behavior – also contains the principal harmonic and multiple high-frequency superharmonics. This means that all the estimated instantaneous functions, such as natural frequency, damping, etc., are oscillating functions. By considering the primary, secondary and other high superharmonics of motion we can identify precisely the initial nonlinear spring and damping force characteristics. To obtain the high superharmonics of the motion, we can use the HVD, which decomposes instantaneous functions into the sum of their separate components. Calculation of the algebraic sum of the corresponding envelopes (11.4) and (11.5) of high harmonics gives the precise initial spring and damping force characteristics.

As an example of the forced vibration regime, let us consider a dynamic system with a combined nonlinear cubic hardening spring and nonlinear dry friction $\ddot{x} + 0.7\text{sgn}(\dot{x}) + x + 0.01x^3 = \sin(\int_0^T \omega_F dt)$; $\omega_F = 2\pi[0.15..0.25]$, $T = 10^4$, where ω_F is the cycle frequency of excitation (radians per second), which is slowly increasing at a constant rate. A simulated forced vibration is shown in Figure 11.9, where a piece of the displacement envelope is shown separately with a zoom to emphasize its oscillating behavior.

Applying the HVD method and considering the first two super harmonics, we are more precisely able to estimate the resultant nonlinear restoring and damping force characteristics of the tested system (11.4) and (11.5) (see Figure 11.10). The fast-varying instantaneous natural frequency (before decomposition), plotted as a fast-changing spiral with a thin line, forms a kind of solid background in Figure 11.10a. The lowpass filtered skeleton curve (Figure 11.10a, dashed line) only approximately restores the nonlinear forces correct to the first term of the motion. The final precise congruent skeleton curve, which also considers two high superharmonics, is shown in Figure 11.10a by a bold line.

It is notable that in the presence of a cubic spring the congruent skeleton curve looks like a tangent to the varying instantaneous natural frequency – mostly for large amplitudes. Contrary to this, in a small amplitude range, the precise congruent skeleton curve crosses the instantaneous natural frequency, mainly due to the influence of the dry friction.

The resultant identified cubic force characteristic (Figure 11.10b, bold line) completely coincides with the characteristic of the initial spring static force $k(x) = 1 + 0.01x^3$ (Figure 11.10b, dashed line). The identified friction force characteristic (Figure 11.10d, bold line) is very close to the initial straight line $h(\dot{x})\dot{x} = 0.07\text{sgn}(\dot{x})$ (Figure 11.10d, dashed line). A least-squares fitting of the identified friction force characteristics to the horizontal polynomial model returns a nonlinear coefficient

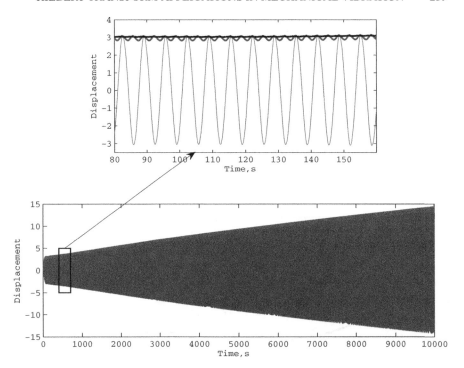

Figure 11.9   Combined nonlinear spring and damping forced vibration: the displacement (—), the envelope (—), the congruent envelope (—)

equal to 0.076, which differs by less than 1% from the initial nonlinear friction coefficient.

The HT identification method in a time domain, as a nonparametric method, is recommended for the identification of instantaneous modal parameters, including the determination of the system skeleton curve (backbone), damping curves, and static force characteristics. The instantaneous modal parameters of nonlinear systems are oscillating functions due to deviations from the linear relationship between specified input and output of the system. These nonlinear distortions are characterized by the appearance of frequencies that are linear combinations of the fundamental frequencies and all the high harmonics presented in the system output signal.

Modern nonstationary vibration decomposition approaches divide the real multicomponent motion into a number of separate principal and other high-frequency superharmonics. By considering high superharmonics, these approaches yield a more precise identification of nonlinear systems, including nonlinear elastic and damping static force characteristics. Theoretically summarizing an infinite number of partial static force characteristics will provide the exact initial static force characteristics of a nonlinear system.

The use of HT methods, based on the nonstationary signal decomposition, are suggested not only to identify nonlinear SDOF systems under free or forced vibration

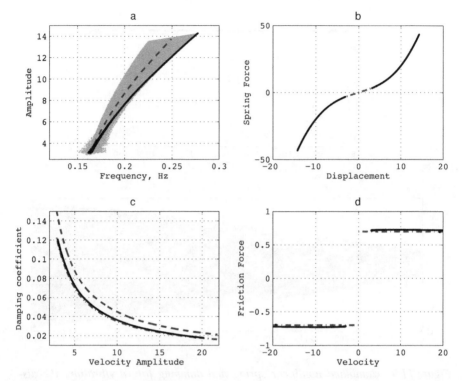

Figure 11.10   The identified parameters of combined nonlinear spring and damping forced vibrations: the skeleton curve (a): instantaneous (—), averaged (- -), identified with high harmonics (—); the spring static force characteristics (b): initial (⋯), identified with high harmonics (—); the damping curve (c): initial (⋯), averaged (- -), identified with high harmonics (—); the friction force characteristics (d): initial (⋯), identified with high harmonics (—)

conditions, but mainly to identify a computer simulation and a precise experiment data that contains detailed information on nonlinear superharmonics. It is important that not only the presence of a nonlinearity is detected, but also that an adequate and readily interpretable dynamic system model is identified.

11.7   Identification of asymmetric nonlinear system

Vibrating systems can have symmetric static force–displacement characteristics (symmetric with respect to the origin) as well as asymmetric characteristics of the nonlinear restoring force as Helmholtz oscillator. The HT identification technique can be simply extended for vibration systems with asymmetric nonlinearities. The main idea of the method is to take the solution of the asymmetric vibration system in the time domain and split it into two different "subsolutions" – separated for positive and negative displacement. Then, the separated solutions of the asymmetric system are determined by adopting the HT method developed for the identification of the nonlinear vibration system (Section 9.2).

11.7.1 Asymmetric nonlinear system representation

The equation of the free vibration of a nonlinear asymmetric structure can be written as follows:

$$\ddot{x} + 2h\dot{x} + k(x) = 0; \quad k(x) = \begin{cases} k_1(x), \text{ if } x > 0 \\ k_2(x), \text{ if } x \leq 0 \end{cases}, \tag{11.6}$$

where x is the system solution, h is the damping coefficient, $k(x)$ is the nonlinear asymmetric spring force, $k_1(x)$ is the nonlinear stiffness characteristic for a positive displacement, and $k_2(x)$ is the nonlinear stiffness characteristic for a negative displacement. The solution of the equation depends mainly on the asymmetric elastic force, which can be expressed in different terms for positive and negative displacements. Thus, in the case of a positive displacement $x > 0$ the solution is produced by the first line of the asymmetric elastic force characteristic (11.6). Each displacement sign changing from positive to negative, or the reverse, switches the vibration structure which will, respectively, include the first or second asymmetric elastic force characteristics. The switching oscillating elastic force will be transformed into an asymmetric oscillatory motion with corresponding amplitude and frequency features.

Let us assume that the solution of a vibration system consists of two independent separate parts: a positive motion, associated only with the positive force characteristic, and a negative motion associated only with the negative force characteristic:

$$x(t) = \begin{cases} x_1(t), \text{ if } x > 0 \\ x_2(t), \text{ if } x \leq 0 \end{cases}. \tag{11.7}$$

The positive part is influenced only by a positive force and, conversely, the negative part of the motion is dependent upon a negative force. In other words, according to this assumption, each part of the solution of the system is determined only by its corresponding force characteristic. Each part of the vibration signal could be represented in the analytic signal form of $x_{1,2}(t) = A_{1,2}(t) \cos \left[\int \omega_{1,2}(t) \right]$ where $x_i(t)$ is the vibration signal (the real-valued function), $A_i(t)$ is the envelope (the instantaneous amplitude), and $\omega_i(t)$ is the IF.

11.7.2 The Hilbert transform identification technique

To separate the positive and negative parts of the signal, and to estimate partial instantaneous characteristics, we will use the previously presented HVD method along with the congruent EOE approach. According to the Hilbert decomposition, the signal can be built up from a slow-varying offset function and several alternate quasiharmonics with varying characteristics. The decomposed congruent quasiharmonic components all together form the EOE function — according to their phase relations. In this way the congruent envelope aggregates all the simplest component envelopes:

$$A_{\text{EOE}}(t) = \sum_{l=1}^{N} A_l(t) \cos \phi_l(t), \tag{11.8}$$

where $A_l(t)$ is the envelope of the l component, and $\phi_l(t)$ is the phase angle between the largest and the l component. For the asymmetric signal: instead of a single congruent envelope, we will construct two envelope functions separately for the signal's positive and negative parts $A_{\text{EOE}}(t) = \begin{cases} A_p, & \text{if } x > 0 \\ A_n, & \text{if } x \leq 0 \end{cases}$. The IF of the asymmetric signal will match the frequency of the sequentially alternating positive and negative parts of the signal: $\omega(t) = \begin{cases} \omega_p(t), & \text{if } x > 0 \\ \omega_n(t), & \text{if } x \leq 0 \end{cases}$.

As a result of the decomposition of the asymmetric signal, we will have two sets of envelopes and instantaneous frequencies, each one for its part of the motion. Each set is suitable for the further HT identification in the time domain. For instance, for a free vibration regime we can use the following formulas (9.4): $h(t) = -\dot{A}/A - \dot{\omega}/2\omega$, $\omega_x^2(t) = \omega^2 - \ddot{A}/A + 2\dot{A}^2/A^2 + \dot{A}\dot{\omega}/A\omega$, where A is the envelope, and ω is the IF of the solution with the first and second derivatives. Here $\omega_x^2(t)$ is the instantaneous modal frequency of the system, and $h(t)$ is the instantaneous modal damping coefficient of the system.

11.7.3   Asymmetric nonlinear system examples

All simulation examples demonstrate the performance that can be achieved using the proposed technique. In order to focus our asymmetric signal processing on the effects of the identification of nonlinear vibration systems, we also use the HT FREEVIB identification method.

11.7.3.1   Asymmetric bilinear system

First, consider the case of a vibration conservative system with an asymmetric bilinear force elastic characteristic $k_{1,2} = \omega_{1,2}^2 x$: $\ddot{x} + 2\dot{x} + k(x) = 0$, $k(x) = \begin{cases} (2\pi 10)^2 x, & \text{if } x > 0 \\ (2\pi 20)^2 x, & \text{if } x \leq 0 \end{cases}$. According to assumption (11.6), the solution of the system is built up from two alternate harmonics. During a half of the period, when the displacement is positive, the vibration appears as a harmonic $A_1 \cos \omega_1 t$; during the next half, when the displacement is negative, the vibration continues as another harmonic with a different amplitude and frequency $A_2 \cos \omega_2 t$. The solution of the asymmetric system is shown in Figure 11.11 along with separated envelopes. By applying the HT identification technique to the signal, the backbones and damping curves shown in Figure 11.12 are obtained. It is obvious that the natural frequencies are constant (respectively, 10 and 20 Hz), and that the damping coefficient is also a constant ($h = 1$, s^{-1}).

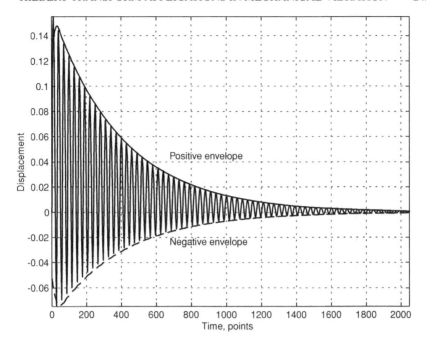

Figure 11.11   The free vibration of the asymmetric bilinear system

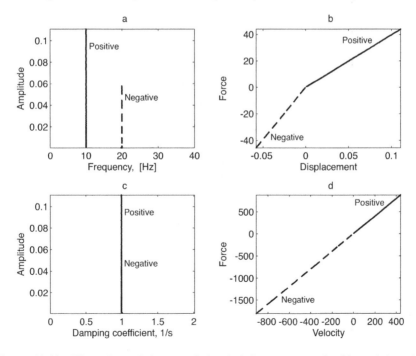

Figure 11.12   The estimated characteristic of a bilinear system: backbone (a), spring force characteristic (b), damping curve (c), damping force characteristics (d)

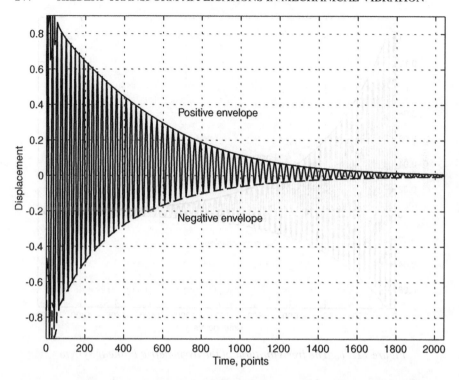

Figure 11.13   The free vibration of the asymmetric system with two cubic stiffnesses
(Feldman, ©2011 by Elsevier)

11.7.3.2   Two different asymmetric cubic stiffnesses

The next example is the case of an asymmetric free vibration with two cubic stiffnesses:

$$\ddot{y} + 2\dot{y} + F(y) = 0, \quad F(y) = \begin{cases} (20\pi)^2(1 + 5y^2)y, \text{ if } y > 0 \\ (40\pi)^2(1 - 3y^2)y, \text{ if } y \le 0 \end{cases}. \tag{11.9}$$

A computer simulation, performed for 1024 points with a sample frequency of 300 Hz, is shown in Figure 11.13. The two obtained backbones shown in Figure 11.14 indicate the varying value of the system frequency. The hardening backbone for a positive displacement and the softening backbone for a negative displacement agree with the initial asymmetric Duffing equation (11.9).

11.8   Experimental identification of a crack

A simple rotor test rig was built in order to apply the proposed procedure to real measurements (Feldman and Seibold, 1999). The test rig consists of a shaft with a radius $R = 9$ mm on hinged supports. A disk is mounted in the middle. Initiated

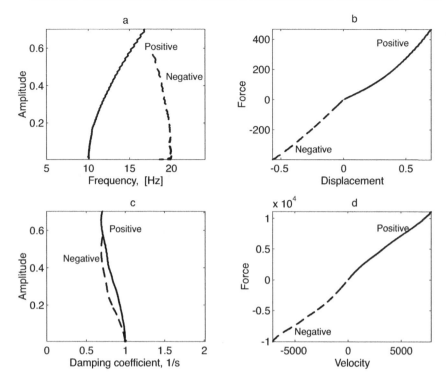

Figure 11.14   The estimated characteristic of a system with two cubic stiffnesses: backbone (a), spring force characteristic (b), damping curve (c), damping force characteristic (d) (Feldman, ©2011 by Elsevier)

by a notch of about 2 mm depth, a transverse crack is introduced in the shaft at a location close to the disk. Two kinds of vibrations are measured: free vibrations of the disk initiated by an impulse, and stationary vibrations at constant speed. Then, static overloads are applied using a special apparatus. This results in dark lines (beach marks) on the crack face. After the experiment, the measurements taken can be related to the beach marks and to the actual crack depths. In this way, the results of the identification can be checked. An experimental identification of the rotor structure with a notch and a crack was made on the basis of four separate measurements (512 time steps) of free vibrations of the disk (a crack at the lowest position).

In this experiment, the depth of the crack was 5 mm. Free vibrations were picked up from the nonrotating rotor after exciting the system with an impulse hammer. The proposed HT method made it possible to separate the measured asymmetric vibration into two different parts. The first part (Figure 11.15, bold line) describes the system behavior for only a positive displacement, and the second part (Figure 11.15, dashed line) describes the behavior for only a negative displacement. The obtained results of the HT identification, shown in Figure 11.15, indicate the closely-spaced positive and corresponding negative backbones of the four separate measurements. The tested system has a strong asymmetric elastic force characteristic: the positive movement

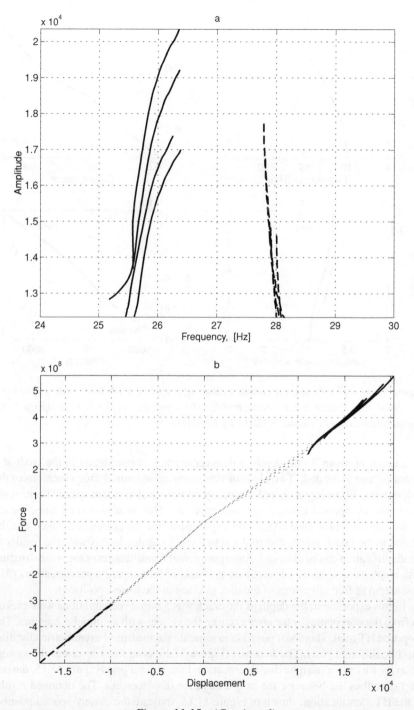

Figure 11.15   (Continued)

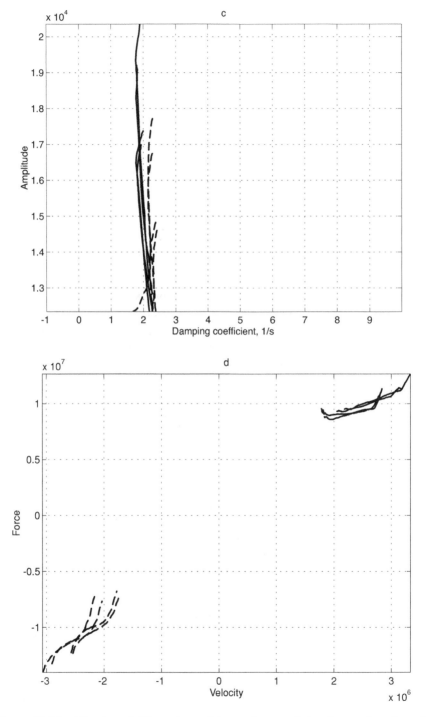

Figure 11.15   (Continued) The estimated force characteristic of a crack and notch structure: backbone (a), spring force characteristic (b), damping curve (c), damping force characteristic (d)

exhibits smaller values of a natural frequency (25.5–26 Hz). The difference between the natural frequencies for positive and negative movements is about 2 Hz.

The system also has a small nonlinear elastic force characteristic, which is dissimilar for positive and negative movements. The positive movement backbone looks like a hardening nonlinear spring. On the other hand, the negative movement backbone shows the characteristics of a small softening nonlinear spring. The tested system has a symmetric friction characteristic. The estimated friction force characteristics possess a strong dry friction nature.

These results clearly indicate a presence of a crack. Now, the location and depth of this crack can be determined by employing the multi-hypothesis approach described in Feldman and Seibold (1999), and the measured vibrations of the disk at a constant speed. It was shown that the crack depth can be identified on the basis of a nonlinear model and, furthermore, even smaller cracks can be diagnosed very well.

11.9   Identification of MDOF vibration system

There are a variety of methods for identifying and analysing multi-degree-of-freedom (MDOF) nonlinear oscillators (Kerschen *et al.*, 2006). Most researchers recognize that a proper signal characterization of nonlinearities is sufficient for the identification of a system. Thus, Ta and Lardies (2006) propose a continuous wavelet transform for identifying and quantifying the nonlinearities of each vibration mode; while the approach in Elizalde and Imregun (2006) is based on frequency response functions and first-order describing functions, which represent the nonlinearities as amplitude-dependent coefficients.

Yang *et al.* (2003a, 2003b) and Poon and Chang (2007) proposed that the EMD method be used to identify normal and complex modes of linear and nonlinear MDOF systems using free vibration responses. Really, the EMD method can decompose a signal into a superposition of intrinsic mode functions that contain one major mode at any time instant. It was first shown that the IMF components of the displacement and velocity responses of a nonlinear elastic structure are numerically close to the nonlinear normal mode responses (Poon and Chang, 2007). The research works of Kerschen *et al.*(2008) and Pai and Hu (2006) also used the popular EMD of a vibration signal for a nonlinear identification. In the majority of cases an *a priori* spatial nonlinear model with a preliminary solution analysis is required for the proper identification of a system. For example, Raman, Yim, and Palo (2005) use an approximate analytical reverse multi-input/single-output technique for studying how nonlinear equations of motion are governed.

In principle, when a nonlinear detailed model structure and a solution are known, the identification can be turned into a parametric identification problem, where unknown parameters can be derived just by fitting the algebraic expression to the data. Those parametric identification techniques, based on an *a priori* nonlinear model, are suitable for chosen models only. Nonparametric identification methods – such as that by Bellizzi, Guillemain, and Kronland-Martinet (2001) do not require an *a priori* model – are preferable. Can such a nonparametric identification approach make it possible to restore different nonlinear multiple-coupled oscillators? Such an approach should be able to detect and characterize the type and degree of nonlinearities, and the unknown vibration model structure and parameters, by considering only the measured vibration and excitation data.

The HT nonparametric methods, FREEVIB and FORCEVIB, which were developed for the nonlinear identification of SDOF systems, are not dedicated directly to MDOF systems, because the MDOF systems produce solutions with multiple components. If, during the first stage, we could separate this vibration composition and get a set of monocomponent signals, would the identified modal parameters show the same nonlinearities as they do in the initial coupled MDOF spatial system? It is not so obvious. Naturally, the MDOF systems are unique due to the coupling, which forms relations and ties together the vibration motions of the linked spatial subsystems. Indeed, if a MDOF system is just formed by uncoupled subsystems, described for example by a modal model with independent coordinates, the nonlinear normal modes can be identified separately as the simple summation of SDOF systems.

It is well known that coupling moves the modal resonance frequencies away from the initial spatial natural frequencies. So the question is: Does the coupling also change nonlinear skeleton curves of the initial spatial subsystems?

The goal of this section is to clarify some specific features of the couplings, examine their influence on the free vibration of nonlinear multiple-coupled oscillators, and explore the available engineering options for an identification of the initial nonlinear spatial mathematical model of MDOF systems (Feldman, 2007b). The research focuses on the common phenomena of coupled nonlinear oscillators with typical linear and nonlinear coupling structures.

11.9.1   Identification of linear coupled oscillators

An analytical vibration analysis in mechanical engineering deals mainly with second-order equations of motion. If these equations are combined into a system they constitute an initial mathematical spatial model with proper physical parameters (mass, stiffness, and damping properties) describing the dynamic behavior of the test structure. Some typical engineering structures also contain localized nonlinearities (joints, geometric discontinuities, shock absorbers, etc.). The primary objective of an identification process is to derive the initial spatial mathematical vibration model, including coupling characteristics and significant nonlinear elements. For reasons of simplicity we consider a system with only two coupled vibration equations, which completely describe the main properties of MDOF systems. The system is assumed to exhibit normal modes, so the viscous damping matrix can be formed by the inverse modal transformation method.

11.9.2   Spring coupling

For example, the 2DOF linear system includes two linked equations of a free motion:

$$\ddot{\varphi} + 2h_\varphi \dot{\varphi} + \omega_\varphi^2 \varphi - \eta_\varphi \xi = 0$$
$$\ddot{\xi} + 2h_\xi \dot{\xi} + \omega_\xi^2 \xi - \eta_\xi \varphi = 0 \tag{11.10}$$

where φ and ξ are coordinates of a coupled vibration motion; ω_φ^2 and ω_ξ^2 are spatial (partial) frequency squares, equal to a natural frequency square of the uncoupled (separated, single) equation; h_φ and h_ξ are damping coefficients; and η_φ and η_ξ are coupling spring coefficients. Damping is assumed to be small, and thus linear proportional damping can be employed, represented here by linear modal damping coefficients. Relations between every spatial frequency and its corresponding vibration amplitude specify the spatial skeleton curve, which in the case of linear oscillators are trivial straight lines.

For further analysis we use substitutions in the form of the analytical signal: $X = x + j\tilde{x}$; $\dot{x} = X\left(\dot{A}/A + j\omega\right)$; $\ddot{X} = X\left(\ddot{A}/A - \omega^2 + 2j\dot{A}\omega/A + j\dot{\omega}\right)$, where $\tilde{x}$ is the HT of x; A, $\dot{A}$, $\ddot{A}$ are the signal envelope and its derivatives; and ω, $\dot{\omega}$ are the signal IF and its derivative. In reality, damping coefficients and derivatives of the envelope are much less than natural frequencies, so their influence can be ignored ($\dot{A}/A = \ddot{A}/A = 0$). As a result, the real part of (11.10) will get a set of coupled equations

$$\varphi\left(-\omega^2 + \omega_\varphi^2\right) - \eta_\varphi\xi = 0$$
$$\xi\left(-\omega^2 + \omega_\xi^2\right) - \eta_\xi\varphi = 0 \tag{11.11}$$

whose determinant is equal to zero, $\omega^4 - \omega^2\left(\omega_\varphi^2 + \omega_\xi^2\right) + \omega_\varphi^2\omega_\xi^2 - \eta_\varphi\eta_\xi = 0$, which produces a known biquadratic equation for the calculation of the modal normal frequencies:

$$\omega_{1,2}^2 = \frac{1}{2}\left\{\omega_\varphi^2 + \omega_\xi^2 \pm \left[\left(\omega_\varphi^2 - \omega_\xi^2\right)^2 + 4\eta_\varphi\eta_\xi\right]^{1/2}\right\} \tag{11.12}$$

The relations between the modal frequencies versus the initial spatial frequencies (11.12), known as Wien's graph, illustrate the fact that the initial spatial frequencies always lie between the modal frequencies (Migulin et al., 1983). So each natural frequency differs from the initial spatial (partial) subsystem natural frequency. In other words, the modal frequencies obtained are not those of the individual component systems. The difference between the spatial and modal frequencies, controlled by coupling coefficients, is governed by the following decoupling coordinate transformation.

Each equation of the system (11.11) characterizes the coefficient of the amplitude ratio (distribution) ψ between the oscillations of different coordinates at every modal frequency:

$$\left(\xi/\varphi\right)_1 = \left(\omega_\varphi^2 - \omega_1^2\right)/\eta_\varphi = \psi_1; \quad \left(\varphi/\xi\right)_2 = \left(\omega_\varphi^2 - \omega_2^2\right)/\eta_\varphi = \psi_2. \tag{11.13}$$

These relative amplitudes, which correspond to every modal frequency – known as the mode shapes – are also the fundamental inherent properties of a freely vibrating, undamped MDOF system. The process of calculating modal parameters from the initial spatial system refers to an analytical modal analysis, which transforms, or decouples, the spatial system into a system of several equations, one for each single mode of vibration. This means that the calculated modes of a vibration effectively uncouple the dynamic equations of motion according to the following relationships between the initial coordinates φ, ξ and the normal coordinates x, y (Migulin *et al.*, 1983):

$$\varphi = x + y, \quad \xi = \psi_1 x + \psi_2 y; \quad x = (\xi - \varphi\psi_2)/(\psi_1 - \psi_2), \quad y = (\psi_1\varphi - \xi)/(\psi_1 - \psi_2). \tag{11.14}$$

The initial coordinate of every DOF is often a real physical coordinate (direction) of vibration measurement, whereas the normal coordinate is a virtual (abstract) coordinate. The real oscillations of the masses can be written as linear combinations of the normal modes.

Expressions (11.14) show that a coupling, as an important common property, completely defines the relations between modal and spatial parameters. The coupling also defines the energy exchange and the time of energy transfer between the participating subsystems. As the coupling strength in coupled subsystems is increased from zero, the oscillations affect each other more and more. Thus the motion generated from one coordinate will appear stronger in the vibration of other coupled coordinates. Thus a coupling gives rise to a multiplicity of natural oscillations in an observed vibration motion. The coupling strength coefficient σ characterizes the degree of coupling between two subsystems (Migulin *et al.*, 1983): $\sigma = 2\sqrt{\eta_\varphi \eta_\xi} / |\omega_\varphi^2 - \omega_\xi^2|$. Notice that a damping coupling, no more than a small partial damping, has almost no influence on the coupled vibration, natural frequencies, or modal shapes.

11.9.3   Reconstruction of coupling coefficients

Traditionally the model in question is only a modal model because modal properties of the system most closely describe the dynamic behavior observed in the tests. For purposes of the identification of MDOF systems, this is not sufficient. A modal test based entirely on a measured vibration data should lead to further inverse reconstruction of a spatial coupled mathematical model based on mass, stiffness, damping properties, and coupling stiffness forces, which have a physical meaning. During the vibration test, modal modes are excited, enabling modal frequencies ω_i and mode shapes ψ_i to be observed and estimated. It is enough to reconstruct the initial partial natural frequencies and the spring couplings of the spatial model. In effect, two equations from (11.12) and two from (11.13) together involve only four unknowns; the general case of n DOF yields $2n$ linear equations with $2n$ unknowns. Therefore, the direct solution of the linear system returns the initial spatial model parameters:

$$\omega_\varphi^2 = (\psi_1\omega_2^2 - \psi_2\omega_1^2)/(\psi_1 - \psi_2); \quad \omega_\xi^2 = (\psi_1\omega_1^2 - \psi_2\omega_2^2)/(\psi_1 - \psi_2)$$
$$\eta_\varphi = (\omega_2^2 - \omega_1^2)/(\psi_1 - \psi_2); \quad \eta_\xi = \psi_1\psi_2(\omega_1^2 - \omega_2^2)/(\psi_1 - \psi_2) \tag{11.15}$$

where ω_φ^2 and ω_ξ^2 are resultant spatial frequency squares, ψ_φ and ψ_ξ are measured mode shapes (amplitude ratios), and ω_1^2 and ω_2^2 are measured modal frequency squares. The above formulas allow us to derive an initial mathematical spatial model to describe the dynamic behavior of the test system without an *a priori* model description.

11.10   Identification of weakly nonlinear coupled oscillators

Typically, every nonlinear equation expressing a vibration motion has a fixed structure. The structure classically includes three independent elements: a restoring elastics force (stiffness, spring) as a nonlinear function of the displacement (position), a damping force (friction) as a nonlinear function of velocity (the first derivative of the position with respect to time), and an inertial force proportional to the acceleration (the second derivative of the position with respect to time). Every independent restoring and damping force element is an *a priori* unknown nonlinear function of motion, as, for example, a hardening or softening spring or a dry or turbulent friction.

The nonparametric identification of nonlinear vibration oscillators – as a typical dynamics inverse problem – deals with *a priori* unknown nonlinear restoring and damping functions. The investigated vibration system with unknown restoring and damping forces moves under (or without) an excitation force. By observation (experiment), we acquire knowledge of the position and/or velocity of the object, as well as the excitation at several known instants of time. The nonparametric identification will determine the initial nonlinear restoring and damping forces. In the case of a free vibration we have only an output signal – the vibration of oscillators; in the case of a forced vibration we deal also with the input excitation.

Nonlinear vibration MDOF systems can consist of essential nonlinear oscillators joined with linear couplings, or of linear oscillators coupled with essential nonlinear attachments. They could also represent a united case of both nonlinear oscillators and nonlinear couplings acting in combination.

11.10.1   Coupled nonlinear oscillators with linear coupling

In general, nonlinear systems composed of several masses, nonlinear springs, and dampers require a more complicated representation. First, let us consider the equations of motion for a coupled 2DOF system, wherein the stiffness nonlinearity depends only on the displacement:

$$\ddot{\varphi} + \omega_\varphi^2 \varphi + \alpha \varphi^3 - \eta_\varphi \xi = 0$$
$$\ddot{\xi} + \omega_\xi^2 \xi - \eta_\xi \varphi = 0 \tag{11.16}$$

Here the first equation includes a simple, relatively weak nonlinear cubic stiffness $\alpha \varphi^3$ corresponding to the initial linearized spatial skeleton curve (8.9) $\omega_0^2(A) = \omega_\varphi^2 + {}^3/_{4\alpha} A^2$. This model combines weakly nonlinear oscillators (without bifurcations, jumps, and chaotic behavior) that have a slow-varying solution in the time domain.

Again, for further analysis, we will use substitutions in the form of an analytical signal considering the overlapping spectra property of the HT of nonlinear functions. The HT substitution establishes direct relationships between the initial parameters of the differential equations and the instantaneous amplitude and frequency of the vibration response. The HT reduction permits the direct construction of an approximate solution defined as a single quasiharmonic with a slow-varying amplitude and frequency. In essence, the HT approach is just an alternative to some well-known linearization methods, such as the harmonic balance linearization. The HT of the harmonic cube will contain the first and third components. Hence, neglecting high-tripled frequency yields: $H[A\cos^3\theta] = A^3(3\sin\theta + \sin 3\theta)/4 \approx \frac{3}{4}A^3 3\sin\theta-$ (see Section 2.4).

The real part of (11.16) takes the form of:
$$\begin{aligned}\varphi\left(-\omega^2 + \omega_\varphi^2 + \frac{3}{4}\alpha A_\varphi^2\right) - \eta_\varphi \xi &= 0 \\ \xi\left(-\omega^2 + \omega_\xi^2\right) - \eta_\xi \varphi &= 0\end{aligned}.$$

We will not solve the obtained nonlinear system, but only analyze the corresponding biquadratic equation for the fixed amplitude $\omega^4 - \omega^2\left(\omega_\varphi^2 + \omega_\xi^2 + \frac{3}{4}\alpha A_\varphi^2\right) + \omega_\varphi^2\omega_\xi^2 + \frac{3}{4}\alpha A_\varphi^2\omega_\xi^2 - \eta_\varphi\eta_\xi = 0$. This gives the modal frequencies $\omega_{1,2}(A)$ of the normalized oscillations as functions of the amplitude:

$$\omega_{1,2}^2(A) = \frac{1}{2}\Bigg\{\omega_\varphi^2 + \omega_\xi^2 + \frac{3}{4}\alpha A_\varphi^2 \pm$$

$$\left[\left(\omega_\varphi^2 - \omega_\xi^2\right)^2 + 4\eta_\varphi\eta_\xi + \frac{9}{16}\alpha^2 A_\varphi^4 + \frac{3}{2}\alpha A_\varphi^2(\omega_\varphi^2 - \omega_\xi^2)\right]^{1/2}\Bigg\} \quad (11.17)$$

These varying modal natural frequencies of the coupled subsystems are also fundamental properties, known as modal skeleton curves; they are independent of the choice of coordinates or any external excitation. But, again, the modal and the corresponding spatial skeleton curves differ from each other.

The modal skeleton curves obtained from (11.17) are shown in Figure 11.16. They capture the effect of a nonlinear behavior; when a coupling again pushes apart the spatial skeleton frequencies, the low frequency becomes lower and the high frequency becomes higher. But now the obtained normal frequencies, besides being dependent on the coupling, are also functions of the vibration amplitude, so every normal frequency will form a corresponding nonlinear normal skeleton curve. An analysis of (11.17) shows that a coupling transfers a single spatial nonlinearity over all coupled modal skeleton curves, controlling the strength of their nonlinear behavior. As the coupling coefficient increases from zero, all other coupled skeleton curves will be rearranged from trivial vertical lines to a more and more nonlinear form. This new result from (11.17) means that the coupling smears out (spreads) a single spatial nonlinear effect over all nonlinear modal coordinates.

A further analysis of (11.17) shows that the initial nonlinearity influences, concurrently, the entire normal skeleton curves with the same tendency. For example, an initial hardening stiffness will also appear as a hardening in every normal skeleton curve, and a softening will exhibit a softening. But each normal skeleton will only be qualitatively similar in appearance to the initial spatial skeleton curve. However, the normal and initial spatial skeleton curves will differ quantitatively from each other. To identify an initial spatial skeleton curve we need to consider the initial spatial frequencies for a nonlinear case. Coefficients of the amplitude distribution in a nonlinear system will also vary as functions of the amplitude: $(\xi/\varphi)_1 = \left(\omega_\varphi^2 - \omega_1^2 + \frac{3}{4}\alpha A_\varphi^2\right)/\eta_\varphi = \psi_1$; $(\xi/\varphi)_2 = \left(\omega_\varphi^2 - \omega_2^2 + \frac{3}{4}\alpha A_\varphi^2\right)/\eta_\varphi = \psi_2$.

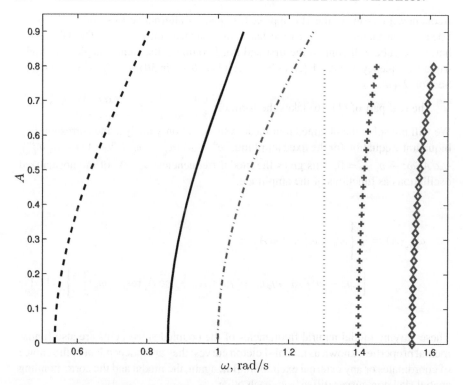

Figure 11.16 Influence of the coupling coefficient η on the skeleton curves of the first ω_1 and the second ω_2 mode of a nonlinear 2DOF vibration system: ω_1, $\eta = 1$ (--); ω_1, $\eta = 0.5$ (—); (---); ω_2, $\eta = 0$ (...); ω_2, $\eta = 0.5$ (+), ω_2, $\eta = 1$ ($\Diamond$)

Such a variation in the amplitude of mode shapes indicates the presence of a nonlinear element, in the initial spatial equation. From coefficients of the amplitude distribution above, and the modal skeleton curves (11.17), we can restore both the coupling coefficients $\eta_{\varphi,\xi}$ and the initial spatial skeleton curves $\omega_{\varphi,\xi}^2(A)$:

$$\omega_{\varphi}^2(A) = \left(\psi_1\omega_2^2 - \psi_2\omega_1^2\right)/(\psi_1 - \psi_2) + {}^3\!/\!_4\alpha A^2; \quad \omega_{\xi}^2(A) = \left(\psi_1\omega_1^2 - \psi_2\omega_2^2\right)/(\psi_1 - \psi_2)$$
$$\eta_{\varphi} = \left(\omega_2^2 - \omega_1^2\right)/(\psi_1 - \psi_2); \quad \eta_{\xi} = \psi_1\psi_2\left(\omega_1^2 - \omega_2^2\right)/(\psi_1 - \psi_2)$$

The spatial skeleton curve from the first equation $\omega_{\varphi}^2(A)$ indeed involves an initial localized nonlinearity with the exact physical characteristics of the nonlinear element ${}^3\!/\!_4\alpha A^2$. As this takes place, the spatial skeleton curve of the second linear equation $\omega_{\xi}^2(A)$ remains linear. This means that the restored spatial skeleton curves will retain their same nonlinear behavior as the initial local model with a nonlinear element; the restored spatial skeleton curves will remain linear as long as the initial local model does not include nonlinearities.

11.10.2   Coupled linear oscillators with nonlinear coupling

Consider another MDOF nonlinear system, now governed by two linear oscillators coupled with a nonlinear element:

$$\ddot{\varphi} + \omega_\varphi^2 \varphi - \eta_\varphi \xi - \beta \xi^3 = 0$$
$$\ddot{\xi} + \omega_\xi^2 \xi - \eta_\xi \varphi = 0 \tag{11.18}$$

Here the coefficient η_φ represents the linear spring coupling, while the coefficient β characterizes the nonlinear cubic stiffness coupling due to the spatial coordinate ξ.

The first harmonic of the HT of cube function takes the form: $H\left[A\cos^3\theta\right] \approx \frac{3}{4}A^3 \sin\theta$, so the real part of (11.18) can be written as a set of equations:

$$\varphi\left(-\omega^2 + \omega_\varphi^2\right) - \xi\left(\eta_\varphi + \frac{3}{4}\beta A^2\right) = 0$$
$$\xi\left(-\omega^2 + \omega_\xi^2\right) - \eta_\xi \varphi = 0 \tag{11.19}$$

The determinant of the above equation is equal to zero and gives the following modal normal frequencies as functions of amplitude:

$$\omega_{1,2}^2(A) = \frac{1}{2}\left\{\omega_\varphi^2 + \omega_\xi^2 \pm \left[\left(\omega_\varphi^2 - \omega_\xi^2\right)^2 + 4\eta_\varphi \eta_\xi + 3\eta_\xi \beta A^2\right]^{\frac{1}{2}}\right\} \tag{11.20}$$

The obtained modal skeleton curves depend on the amplitude and are nonlinear functions in spite of the linear nature of the oscillators under consideration. The mode shapes derived from (11.19) are also nonlinear functions:

$$(\xi/\varphi)_1 = \left(\omega_\varphi^2 - \omega_1^2\right)/\left(\eta_\varphi + \frac{3}{4}\beta A_\varphi^2\right) = \psi_1;$$
$$(\xi/\varphi)_2 = \left(\omega_\varphi^2 - \omega_2^2\right)/\left(\eta_\varphi + \frac{3}{4}\beta A_\varphi^2\right) = \psi_2. \tag{11.21}$$

Combining four equations from (11.20) and (11.21) we get back the spatial natural frequencies and coupling coefficients:

$$\omega_\varphi^2(A) = \left(\psi_1\omega_2^2 - \psi_2\omega_1^2\right)/(\psi_1 - \psi_2); \quad \omega_\xi^2(A) = \left(\psi_1\omega_1^2 - \psi_2\omega_2^2\right)/(\psi_1 - \psi_2)$$
$$\eta_\varphi(A) = \left(\omega_2^2 - \omega_1^2\right)/(\psi_1 - \psi_2) - \frac{3}{4}\beta A^2; \quad \eta_\xi = \psi_1\psi_2\left(\omega_1^2 - \omega_2^2\right)/(\psi_1 - \psi_2).$$

Notice that the restored coupling $\eta_\varphi(A)$ correctly represents the initial nonlinear coupling function.

In the general case, MDOF vibration systems can have both types of nonlinearities – the oscillator nonlinear stiffness and the nonlinear stiffness coupling – acting simultaneously. In the case of a combined nonlinear spring and nonlinear coupling, the nonlinear modal skeleton curves will have a rather complicated form. Nevertheless, as shown, the restored spatial skeleton curves and coupling coefficients will return the correct initial spatial characteristics of the system.

11.10.3   HT decomposition and analysis

The estimation of spatial model characteristics requires two groups of amplitude/frequency-varying data: nonlinear modal frequencies and nonlinear modal shapes. New nonstationary signal decomposition methods based on the HT (Huang *et al.*, 1998; Feldman, 2006) work as pseudo-adaptive filters in the time–frequency domain and make it possible to obtain the required amplitude/frequency-varying data required for making a further identification.

Generally, the simultaneously decomposed vibration of a coordinate of nonlinear multiple-coupled oscillators will demonstrate two different physical natures of vibration. The first group in a multicomponent oscillation shows just the presence of a partial motion from the coupled subsystems. These coupled components will also exist in linear MDOF systems. The second group in a multicomponent oscillation is associated with intricate nonlinear relationships in restoring and damping functions, which cause high-frequency superharmonics and intermodulation distortions. Actually, a real motion of nonlinear systems contains several main, or principal, quasiharmonic solutions along with an infinite number of multiple high-frequency superharmonics. In the next sections we will consider only the first group of a motion that includes the sum of the primary system solutions of several nonlinear modes.

In the case of multiple-coupled vibration oscillators, the HT signal decomposition takes apart every nonlinear normal mode. It gets rid of the mode mixing phenomenon and attempts to purify and clean every vibration mode signal. The HT decompositions do not require a preliminary bandpass filtering of the signal in order to pick out each mode of interest and reject all the others. After using the decoupling technique, we will have several corresponding decoupled vibration motions.

The idea is to decompose an initial wideband oscillation $x(t)$ into a sum of elementary components with a slow-varying instantaneous amplitude and frequency, so that $x(t) = \sum a_l(t) \cos \left(\int \omega_l(t) dt \right)$, where $a_l(t)$ are instantaneous amplitudes and $\omega_l(t)$ are instantaneous frequencies of the l-component. Thus, the obtained IF of each synchronous component will correspond to a decoupled modal natural frequency, and the instantaneous amplitude ratio will match the nonlinear normal mode shapes – all as functions of time t.

11.10.4   Modal skeleton curve estimation

Every decomposed modal vibration is a solution of a corresponding modal SDOF second-order system having nonlinear elastic (restoring) force characteristics $k(x)$: $\ddot{x} + h_0(\dot{x}) + k(x) = \ddot{x} + h_0(\dot{x}) + \omega_0^2(x)x = 0$. The nonlinear restoring force can be represented as the multiplication of a varying nonlinear natural frequency $\omega_0^2(x)$ and a nonlinear oscillator solution x. The instantaneous undamped modal natural frequency and the instantaneous damping coefficient of the tested oscillator are estimated according to the FREEVIB or FORCEVIB methods (Chapter 10).

In general, FORCEVIB operates with a SDOF system using single input and output signals. Applying it to a MDOF system requires us to consider an existing mode shape and the polarity of the vibration signal. According to a specific mode shape, an output sensor can be located at the point with an in-phase or out-of-phase vibration motion relative to the excitation of the shaker. Such a system can have nodes – the points are not suitable for the analysis, because around the nodes a specific mode has a small amplitude and no stable phase. Also, as MDOF systems can have points related to another resonance mode that is not excited by the chosen frequency, their phase function cannot be used to estimate damping.

11.10.5   Mode shape estimation

In nonlinear oscillators, the envelope of every decomposed nonstationary modal component varies in time, so the varying modal shape of the i mode takes the form $\psi_i(A) = A_{\xi i}/A_{\varphi i}\cos\theta_i$. Here $A_{\xi i}$ is an envelope of the i decomposed modal component of the spatial coordinate ξ, $A_{\varphi i}$ is an envelope of the same frequency modal component of the next spatial coordinate φ, and θ_i is a phase between the modal vibration components ξ_i and φ_i.

The HT decomposition allows us to detect and isolate modal frequencies and modal shapes even in the case of a time-varying amplitude (mode), changing frequency, envelope decay, and phase variation in time for each isolated mode. Thus, the recent achievements in nonstationary signal decomposition (Huang et $al.$, 1998; Feldman, 2006) open a way for a new combined analysis and an identification of both linear and nonlinear MDOF vibration systems. Next we will describe a common identification scheme and consider some examples of an identification of weakly nonlinear multiple vibration oscillators.

11.10.6   Description of the identification scheme

The main idea of the identification is to apply a linear inverse transformation from the modal to the spatial coordinates – for obtaining correct initial spatial nonlinear characteristics. The free vibration of an undamped structure which is assumed to be linear and proximately discretized for n DOF can be described by the spatial equations of motion: $[\mathbf{M}]\{\ddot{\mathbf{y}}\} + [\mathbf{K}]\{\mathbf{y}\} = \mathbf{0}$, where $[\mathbf{M}]$, $[\mathbf{K}]$ and $\{\mathbf{y}\}$ are the matrices of the spatial mass, stiffness, and vector of the displacement. If the number of measured modes is equal to the number of measured coordinates n, the transformation can be written as (Ewins, 1984): $[\mathbf{M}] = [\mathbf{\Phi}]^{-\mathbf{T}}$, $[\mathbf{K}] = [\mathbf{\Phi}]^{-\mathbf{T}}[\lambda_r^2][\mathbf{\Phi}]^{-1}$, where $[\mathbf{\Phi}] = [\mathbf{\psi}][\mathbf{m_r}]^{-1/2}$ are the mass-normalized eigenvectors, $\mathbf{\psi}$ is the mode shape, $\mathbf{m_r}$ is the modal mass, and λ_r^2 is the eigenvalues corresponding to the $natural\ frequency$ $squared$.

These features of normal modes allow us to define normal modes in terms of eigenvectors (or eigenfunctions) and to express them in the system response as a superposition of modal responses. The modal parameters of nonlinear normal modes can vary with the total energy due to their frequency–energy dependency. For example, the nonlinear normal modes of a nonlinear system may have varying modal frequencies dependent on the vibration amplitude. From the decomposed nonlinear modal oscillations, one can obtain information about the amplitude-dependent modal frequencies and damping parameters.

In addition to a reconstruction of the MDOF spatial model, the coordinate transformation makes a reconstruction of the initial nonlinear elements possible. The actual number of nonlinear elements can be small, but their overall nonlinear behavior may be significant. A provided modal analysis based on the HT signal decomposition will allow an estimation of the modal skeleton curves and mode shapes of every nonlinear mode of the coupled vibration oscillator. But these nonlinear modal characteristics are shifted because of the coupling, and, moreover, they differ quantitatively from the initially identified characteristics. Only the use of coordinate transformations, such as algebraic formulas (11.22), allows a reconstruction of the initial spatial nonlinear characteristics. For different nonlinear vibration systems, the mentioned coupling reconstruction is a nonparametric identification technique that does not require an *a priori* model description.

The spring force characteristics $k(x)$ intended for an identification are defined as the multiplication of two phasors: $K = \omega_0^2 X$, where ω_0^2 is the varying nonlinear natural frequency, and X is the displacement of the vibration in signal analytic form. For a complex product, the magnitudes are multiplied and the angles are added, so the following "envelope" expression returns the initial spatial static force characteristics:

$$k(x) = \begin{cases} \omega_0^2(A)A, & x > 0 \\ -\omega_0^2(A)A, & x \langle 0 \end{cases} \; ; \quad h(\dot{x})\dot{x} = \begin{cases} h_0(a_{\dot{x}})a_{\dot{x}}, & \dot{x} > 0 \\ -h_0(a_{\dot{x}})a_{\dot{x}}, & \dot{x} \langle 0 \end{cases} , \text{ where } a_{\dot{x}} \text{ is the envelope}$$

of the velocity. The estimated average natural frequency and the average damping function include the main information about the initial nonlinear elastic and damping characteristics.

The proposed identification is a three-stage method. It includes the following procedures: (a) the multipoint vibration measurements of every mass belonging to a MDOF system; (b) the HT signal decomposition of the measured vibrations into nonlinear normal modes (synchronous vibration components) and an estimation of the corresponding modal skeleton curves and mode shapes; and (c) an estimation of the initial spatial skeleton curves and couplings and a reconstruction of spatial nonlinear differential equations of the initial model. The proposed nonparametric identification method is dedicated primarily to stable weakly nonlinear oscillators with quasi and almost periodic oscillating-like solutions.

11.10.7   Simulation examples

11.10.7.1   Model 1. Nonlinear oscillators with linear coupling

As the first example, consider a system of two coupled nonlinear equations, where the first (driving) oscillator includes a hardening nonlinear spring element and the second (driven) oscillator includes a softening one:

$$\begin{aligned} \ddot{\varphi} + 0.05\dot{\varphi} + \varphi + \varphi^3 - 0.8\xi &= 0, \quad \varphi_0 = 4.0 \\ \ddot{\xi} + 0.05\dot{\xi} + 5.4\xi - 0.5\xi^3 - 0.6\varphi &= 0 \end{aligned}$$. The above system of equations was

numerically solved using the fourth-order Runge–Kutta method with the time step of $\Delta t = 0.25$ s. The signal modulated waveforms from every spatial coordinate are shown in Figure 11.17a; the decomposed normal smooth components – according to the HVD method – are shown in Figure 11.17b. The estimated ratios between the corresponding decomposed normal envelopes as the mode shapes of every nonlinear mode (Figure 11.17b) are plotted in Figure 11.17c. The estimated mode shapes depend on the amplitude, which immediately exhibits the nonlinear behavior of a

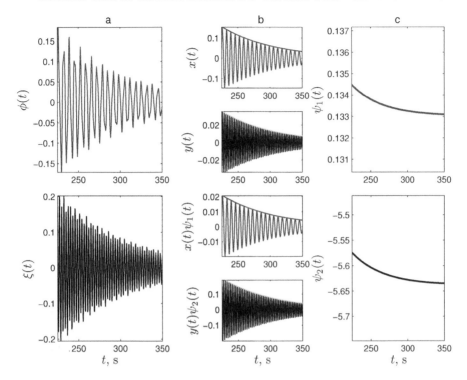

Figure 11.17   The free vibration of two coupled Duffing oscillators: the first $\varphi(t)$ and second $\xi(t)$ spatial coordinate vibration (a), the HT decomposed normal coordinates (b), the mode shapes of two modes (c) (Feldman, ©2011 by Elsevier)

vibration oscillator. In effect, the estimated modal skeleton curves (Figure 11.18a, b, dashed line) are nonlinear, but both the first hardening and the second softening are far away from the initial spatial skeleton curves (Figure 11.18a, b, dotted line). Only the restored spatial skeleton curves (Figure 11.18a, b, bold line) are very close to the initial skeleton curves. In fact, least squares fitting data of the resultant identified spatial skeleton curves (Figure 11.18a, b, bold line) returns nonlinear stiffness coefficient α equal to 0.844 (the initial α for the first mode was equal to 0.75) and to –0.39 (the initial α for the second mode was equal to –0.375), that respectively differs by less than 12% and 4% from the initial nonlinear stiffness coefficient. Both restored coupling static characteristics (Figure 11.18c, bold lines) are close to the initial straight vertical lines (Figure 11.18c, dot-dash lines).

11.10.7.2   Model 2. Oscillators with a nonlinear coupling

The example of a vibration oscillator considered here is a 2DOF system with linear equations of motion and two opposite cubic nonlinear couplings respectively:
$$\ddot{\varphi} + 0.01\dot{\varphi} + \varphi - 0.4\xi - \xi^3 = 0, \varphi_0 = 0.5$$
$$\ddot{\xi} + 0.01\dot{\xi} + 2.88\xi - 0.3\varphi - 0.5\varphi^3 = 0, \xi_0 = 0.7$$. A simulated free vibration is

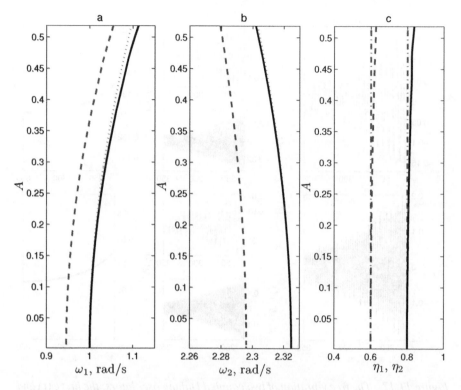

Figure 11.18 The skeleton curves of the first hardening stiffness mode (a): the modal (- -), the identified spatial (—), the initial spatial (···); the skeleton curves of the second softening mode (b): the modal (- -), the identified spatial (—), the initial spatial (···); the coupling of nonlinear modes (c): the first equation identified spatial (- -), the second equation identified spatial (—), the initials (---) (Feldman, ©2011 by Elsevier)

shown in Figure 11.19a. A decoupled modal component after the HVD signal decomposition is illustrated in Figure 11.19b. The corresponding mode shapes (Figure 11.19c) indicate a nonlinear type of motion. Both of the obtained spatial skeleton curves of linear oscillators (Figure 11.20a, b, bold lines) almost coincide with initial vertical straight lines. Every restored coupling static characteristic repeats the corresponding initial cubic spring coupling subsystem (Figure 11.20c). Thus, fitting the least squares data of the characteristics of the nonlinear first mode coupling to a polynomial model returns an nonlinear coefficient β equal to 0.79 while the initial β was equal to 0.75. The estimated nonlinear second mode coupling coefficient β is equal to −0.42 while the initial β for the second mode was equal to −0.375, which differs about 5−10% from the initial nonlinear coefficient values. The comparison between the identified characteristics and the initial ones shows that the proposed approach makes it possible to have a precise estimation of the actual oscillator nonlinearities.

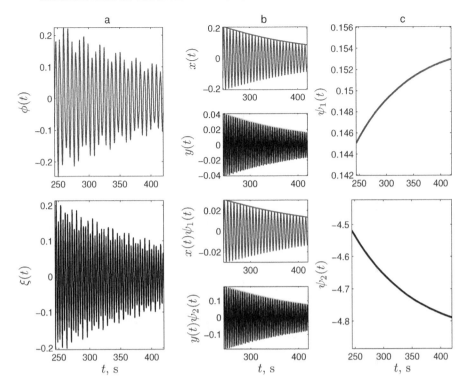

Figure 11.19   The free vibration of two oscillators with nonlinear couplings: the first $\varphi(t)$ and second $\xi(t)$ spatial coordinate vibration (a), the HT decomposed normal coordinates (b), the mode shapes of two modes (c) (Feldman, ©2011 by Elsevier)

11.10.7.3   Model 3. Self-excited coupled oscillators

The next example combines two van der Pol equations with linear coupling:
$$\ddot{\varphi} + 0.2\dot{\varphi}(\varphi^2 - 1) + \varphi - 0.8\xi = 0, \quad \varphi_0 = 0.01$$
$$\ddot{\xi} + 0.3\dot{\xi}(\xi^2 - 1) + 1.88\xi - 0.8\varphi = 0, \quad \xi_0 = 0.02$$
. Every van der Pol equation contains a bilinear cross-term with the multiplication of two variables: the displacement squared and the velocity. Due to the nonzero initial conditions and to the presence of unstable damping terms, both coordinates of the tested model immediately display an increasing self-excited periodic motion (Figure 11.21a). After a transient increasing motion, the observed steady-state solutions include a combination of the self-excited vibrations generated by every coordinate.

Application of the HVD signal decomposition shows these two separated components with different frequencies (Figure 11.21b). The first component $x(t)$ is generated by the first van der Pol equation, and the second component $y(t)$ by the second equation. The amplitude level of both components depends mainly on the initial bounding friction coefficient of the coordinate squared, and not on the coupling coefficients. This means that the estimated ratios between the decomposed normal envelopes (Figure 11.21c) do not describe the regular mode shapes; nevertheless, the existing

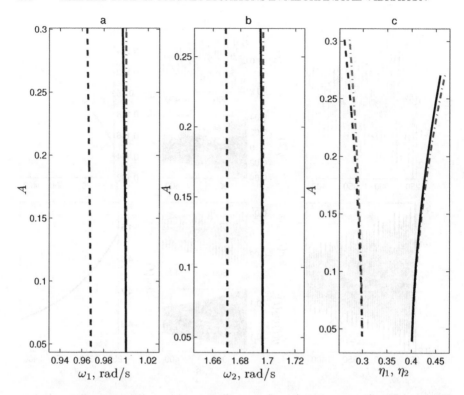

Figure 11.20   The skeleton curves of the first mode with nonlinear couplings (a): the modal (- -), the identified spatial (—), the initial spatial (-□-); the skeleton curves of the second softening mode (b): the modal (- -), the identified spatial (—), the initial spatial (-□-); the coupling of nonlinear modes (c): the first equation identified spatial (- -), the second equation identified spatial (—), the initials (---) (Feldman, ©2011 by Elsevier)

coupling has an impact on the modal skeleton curves, shifting them from their initial spatial frequencies (Figure 11.22ab).

Appling FREEVIB (the HT identification method) to the transient increasing motion of every coordinate will return the initial nonlinear friction force characteristics in the form $\dot{x}(x^2 - 1)$ typical for the van der Pol equation (Figure 11.22c, d).

To summarize our analysis of the representation of the analytic and modeling signal we would like to repeat some conclusions. A single nonlinear element presented only in a single specific equation of a MDOF vibration system will be noticed in the vibration of all coupled coordinates. The resulting behavior is expected, due to the dynamic interaction between coupled vibration subsystems. The observed modal nonlinearity does not correspond to the initial nonlinearity of the spatial model. The observed modal skeleton curve will be similar to the initial spatial skeleton curve in appearance only. However, modal spatial skeleton curves and initial spatial skeleton curves differ quantitatively from each other. To identify the initial spatial

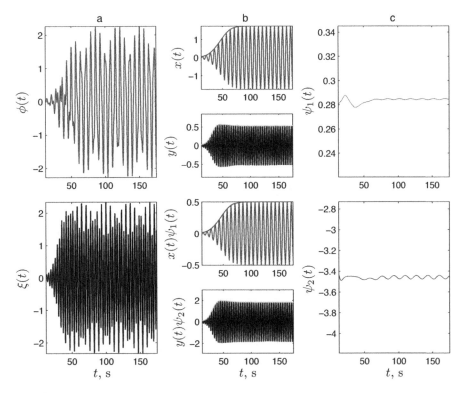

Figure 11.21   Self-excited vibration of two coupled van der Pol oscillators: the first $\varphi(t)$ and second $\xi(t)$ spatial coordinate vibration (a), the HT decomposed normal coordinates (b), mode shapes of two modes (c)

skeleton curve we suggest the application of a modal-spatial coordinate transformation, together with a HT vibration decomposition. The estimated nonlinearities can be actually quantified and included in the dynamics model. The proposed identification method is nonparametric; it does not require *a priori* consideration of a nonlinear nature of the vibration oscillator. The method is recommended for nonlinear parameter identification, including a determination of the system skeleton curve (backbone) and the static coupling force characteristics.

11.11   Conclusions

Modern HT signal decomposition approaches divide the real multicomponent solution into a number of separate principal and high-frequency superharmonics. These approaches, considering the high superharmonics, yield a more precise identification of nonlinear systems, including the nonlinear elastic and damping static force characteristics. Theoretically, the sum of an infinite number of partial static force

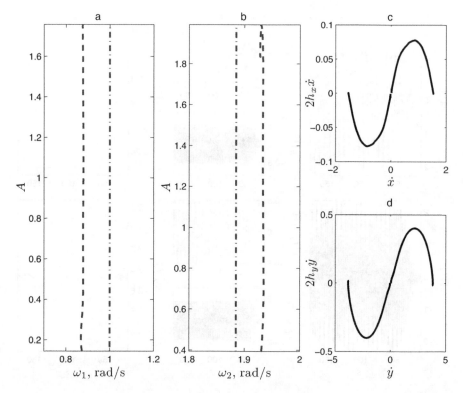

Figure 11.22   Skeleton curves of the first mode (a): the modal (- -), the initial spatial (-□-); skeleton curves of the second mode (b): the modal (- -), the initial spatial (-□-); the identified nonlinear friction force of the first van der Pol oscillator (c); the identified nonlinear friction force of the second van der Pol oscillator (d)

characteristics will provide the exact initial static force characteristics of the nonlinear system.

For nonlinear systems the instantaneous modal parameters obtained after direct HT identification are oscillating functions due to deviations from a linear relationship between the specified input and output signals of the system. In the system output signal these nonlinear distortions are characterized by the appearance of frequencies that are linear combinations of the fundamental frequencies and all the high harmonics. A consideration of these high harmonics can be used for the identification of the congruent modal parameters, including the determination of the precise system skeleton curve (backbone), damping curves, and static force characteristics.

The proposed HT methods, based on nonstationary signal decomposition, are suggested for the identification of nonlinear SDOF systems under free or forced vibration conditions. The proposed HT methods are suggested mainly for an identification of a computer simulation and precise experiment data that contains detail information about nonlinear superharmonics.

When a single nonlinear element is presented in a MDOF vibration system, the resultant nonlinear behavior can be noticed in the vibration of all coupled coordinates. The resulting behavior is expected, owing to the dynamic interaction between coupled vibration subsystems. The observed modal nonlinearity does not correspond to the initial nonlinearity of the spatial model. The observed modal skeleton curve will only be similar in appearance to the initial spatial skeleton curve. However, the modal and the initial spatial skeleton curves differ quantitatively from each other. To identify the initial spatial skeleton curve we suggest the application of a the modal-spatial coordinate transformation together with a HT vibration decomposition. The estimated nonlinearities can actually be quantified and included in the dynamics model. The proposed identification method is a nonparametric method, which does not require *a priori* consideration of the nonlinear nature of the vibration oscillator. The method is recommended for nonlinear parameter identification, including the determination of the system skeleton curve (backbone) and the static coupling force characteristics.

12

Experience in the practice of system analysis and industrial application

We have already considered some examples of the HT application in structural health monitoring and damage detection (see Section 7.1). These applications are mostly based on the extracted features of the level and frequency content of a varying diagnostic signal in the form of the envelope, the IF, and the Hilbert spectrum. In some cases a more sophisticated analysis, based on a physical equation of the dynamic system behavior, may be required (Messina, 2009). The proper recognition of a technical state of a considered object is possible by taking into account additional information about the dynamic system and its vibration model. A dynamic system model is commonly created on the basis of knowledge of a mechanical construction and the functional operation of a technical object. For example, it is only possible to recognize the technical state of a structure if we have information about its modal parameters, such as frequencies, damping, and mode shapes as functions of the physical inertia, stiffness, damping properties. Consequently, changes in physical properties, such as a reduction in stiffness as a result of the onset of cracks or a loosening connection, will cause changes in the modal properties of the system.

In this final chapter some industrial applications of the HT and EMD are considered. Several of these applications are of considerable significance and could easily be extended by including further examples of applied research in mechanical engineering. We have restricted our consideration only to the main specific types of a successful implementation of the HT in practice.

Hilbert Transform Applications in Mechanical Vibration, First Edition. Michael Feldman.
© 2011 John Wiley & Sons, Ltd. Published 2011 by John Wiley & Sons, Ltd.

12.1   Non-parametric identification of nonlinear mechanical vibration systems

At each instant of time, an equation of the system vibration $[M]\{\ddot{x}\} + C\{\dot{x}\} + [K]\{x\} = \{z(t)\}$ links three vector and matrix quantities: a displacement solution $\{x\}$, system matrices of the mass, damping, and stiffness $[M]$, $[C]$, $[K]$, and an external excitation $\{z(t)\}$. It is obvious that one matrix can be determined if the other two are given. Determining the solution from given system parameters and type of excitation is known as a direct problem in differential solution theory (Plakhtienko, 2000). The inverse problem is an identification of unknown system parameters when the excitation input and the output solution are known (Figure 12.1).

Nonlinear behaviors of a dynamic system response in modern industrial designs were considered. For example, a nonlinear friction has been used to enhance engine blade damping, and wire-mesh-bearing dampers have been designed to improve the dynamic stability of rotors. Using the HT achievements for a nonlinear system

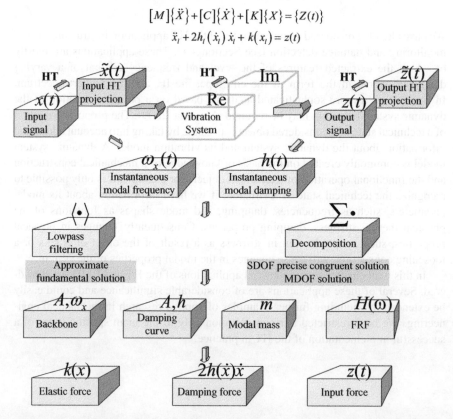

Figure 12.1   HT procedures for the identification of vibration systems (Feldman, ©2011 by Elsevier).

characterization became a very actual and challenging topic for industrial systems (Luo, Fang, and Ertas, 2009; Fang, Luo and Tang, 2008). In addition to the identification of the system's modal parameters, the HT approach allows the non-parametric identification of specific nonlinear elements, such as backlash (Tjahjowidodo, Al-Bender, and Van Brussel 2007), or a nonlinear damping in fluid elastic (Eret and Meskell, 2008), power systems (Laila *et al.*, 2009), and frame structures (Curadelli *et al.*, 2008). In Liu, Messina and Vittal (2004) the EMD method was used in a complementary manner to efficiently and accurately represent the nonlinear modal interaction and distribution of a nonlinear and nonstationary behavior. The EMD and the Hilbert spectra were applied to measure the time series from aero-elastic systems (Lee *et al.*, 2010). The measured nonstationary and nonlinear time series were decomposed into a set of intrinsic oscillatory functions and – from ground and flight tests – used for the identification of aerospace systems by an estimation of modal parameters and a model improvement of aerospace structures.

In some practical situations, it is desirable to estimate the modal parameters in real time. A continuous estimation of these parameters can be used, for example, for classification and diagnostics purposes (Jones, 2009).

12.2 Parametric identification of nonlinear mechanical vibrating systems

When a nonlinear detailed model structure and its solution are known, the identification can be turned into a parametric identification problem, where unknown parameters can be derived just from the measured data. In a parametric identification, mathematical models of the system are assumed to be known with some number of parameters. Sets of ordinary differential second-order equations are taken as mathematical models of identified systems. Those parametric identification techniques, based on an *a priori* nonlinear model, are suitable for the chosen models only.

For example, equations of motion for a slow-flow model – as a 2DOF system with a cubic nonlinearity – were obtained by using a complexification-averaging technique (Kerschen *et al.*, 2008). A proper investigation of the solution decomposed into dominant components by the application of the EMD approach allows us to identify nonlinear system equations governing the amplitude and phase variations in time. Another example of a nonlinear vibration characterization by a signal decomposition considers a number of oscillator models such as the damped Duffing with cubic, quadratic, higher-order stiffness nonlinearities, or a multimode modal coupling subjected to a harmonic excitation (Pai, 2007; Pai and Palazotto 2008; Pai *et al.*, 2008). Performing the EMD of a time-varying solution of the known structure allows us to identify the initial nonlinear model. In a similar manner, Coulomb and quadratic damping can be estimated from the envelope of the response – according to the damping identification techniques based on a simple curve fitting (Smith and Wereley, 1999).

The experimental dynamic response data was processed using the HT technique in Franchetti and Modena (2009). The study focused on a structural linear and nonlinear damping and its relation to crack damage. The HT method used data history in a narrow frequency bandwidth after applying a bandpass filter for each of the SDOF subsystems. After uncoupling, the free response damping coefficient was calculated by an exponential fit with the analytical signal presentation. The proposed method is suited for quality control, and can be further extended for the in situ detection of damage in concrete structures under an ambient vibration.

For the assessment of modal parameters, the Hilbert analysis offers an alternative approach to the more standard parametric techniques (Browne *et al.*, 2009). Practical considerations, such as window selection and zero padding, have been shown to have a significant influence on the accuracy of results. The Hilbert analysis has demonstrated its ability to identify modal parameters in many of the test signals examined, especially where noise conditions are not excessive and modal frequencies are well separated.

The HT has proved to be useful for a long time series; low-dimensional chaotic systems that exhibit transient chaos, consequently give only a short time series (Lai and Ye, 2003). A Hilbert analysis also has the advantage of being suitable for a non-stationary time series to assess the IF spectrum of the system.

12.3   Structural health monitoring and damage detection

When measured data contains damage events or information on a damage in the structure, it is important to extract as much of this information as possible. The HT is not only able to detect the presence of a defect in the structure, but can also quantify the extent of the damage (Feldman, 2009a). Thus, the modal-based methods can quantitatively identify structural system properties before and after damage events, including the modal mass, stiffness, and damping matrices. Several recent applications of the HT methods to structural dynamics are devoted to damage detection by using experimentally measured mode shapes and natural frequencies.

12.3.1   Damage detection in structures and buildings

The EMD and Hilbert spectra are capable of identifying natural frequencies, damping ratios, mode shapes, the stiffness matrix, and the damping matrix of a structure on the base of the measured acceleration responses (Li, Deng, and Dai, 2007). Damage of the benchmark structure has been identified by a comparison of the stiffness in each story prior to and after the damage event (Lin, Yang, and Zhou, 2005).

For some applications dealing with damped vibration, the modal parameters were estimated from the instantaneous phase and the envelope (see Yang, Kagoo, and Lei, 2000). These modal parameters were calculated from the slope of the linear least-squares straight line of instantaneous time functions. The presented method identifies the natural frequencies and damping ratios in situ of tall buildings using ambient wind vibration data. The EMD method is used to obtain general modal responses; the random decrement technique is used to extract the decay free vibration of each modal response. The HT is then used to identify complex eigenvalues, including the natural frequency and damping ratio of each mode.

In Yang *et al.* (2004) the EMD is intended to extract damage spikes due to a sudden change of structural stiffness from the measured data, thereby detecting the instants of time and locations of the damage. A combination of the EMD and the HT is capable of detecting the time instants of damage and can determine the natural frequencies and damping ratios of the structure before and after damage. The methods are applied to a benchmark problem established by the American Society of Civil Engineers Task Group on structural health monitoring. Simulation results demonstrate that the proposed methods provide new and useful tools for detecting and evaluating the damage in structures.

To improve the EMD, Chen and Feng (2003) propose a technique – based on waves beating phenomena – for decomposing components in narrowband signals, where the time-scale structure of the signal is unveiled by the HT as a result of wave beating. The order of component extraction is reversed – from that in the EMD – and the end effect is confined. This technique is verified by performing a component decomposition of a simulated signal and a free decay signal actually measured in an instrumented bridge structure.

12.3.2 Detecting anomalies in beams and plates

A discussion on locating an anomaly – in the form of a crack, delamination, stiffness loss or boundary in beams and plates – can be found in Quek, Tua, and Wang (2003). The results indicate that the EMD is able to give a good representation of a localized event and is sensitive to a slight distortion in the signal. A crack and a delamination in homogeneous beams can be located accurately, and damage in a reinforced concrete slab can be identified if it has been previously loaded beyond the first crack. Sometimes the sensitivity of the EMD is such that an analysis with a distorted signal needs a careful interpretation, as illustrated by an example with an aluminum plate.

In their paper, Pines and Salvino (2006) present an experimental validation of the EMD approach using a civil building model. Empirically derived basis functions are processed through the EMD to obtain magnitude, phase, and damping information. This information is later processed to extract the underlying incident energy propagating through the structure. The damping or loss factor values can be calculated from the response data directly using the Hilbert Damping Spectrum (Li, Deng, and Dai, 2007). The main step is to define a time-dependent decay factor for each empirical mode by the envelope function.

In Douka and Hadjileontiadis (2005) the dynamic behavior of a cantilever beam with a breathing crack is investigated both theoretically and experimentally. The response data is analyzed by applying the EMD and the HT for the IF estimation. It is shown that the IF oscillates between the frequencies corresponding to open and closed states, revealing the physical process of crack breathing. The variation of the IF follows definite trends and therefore can be used as an indicator of the crack size. It provides an efficient and accurate description of the nonlinearities caused by the presence of a breathing crack.

In parallel with the HT, another method – the Wavelets transform – is developed in the signal processing allowing similar applied problems to be proved (Cohen,

1989). Numerous scientific works evaluating these methods are based exclusively on an empirical data comparison. For example, in Kijewski-Correa and Kareem (2006) these two approaches provided comparable evidence of the nonstationary and nonlinear behavior of vibration systems. Given that both transforms can represent nonlinear characteristics, albeit differently, the selection of one approach over the other depends entirely on the perspective desired.

12.3.3   Health monitoring in power systems and rotors

The results of a study of subsynchronous torsional oscillations in power systems with flexible AC transmission system controllers are presented in Andrade *et al.* (2004). An application of an HT-based signal and system analysis techniques indicates that nonlinear oscillations may involve an interaction between fundamental frequencies. These interactions result in a significant modulation of primary frequencies and lead to nonlinear and non-stationary behavior. In the paper a harmonic generation and a nonlinear mode interaction in power systems are detected. The efficacy of the EMD method for the separation of closely spaced modal components is demonstrated in both synthetic and transient stability data (Laila, Messina, and Pal, 2009). It is shown that the method produces a physically motivated basis suitable for the analysis of general nonlinear and nonstationary signals, particularly for the inter-area oscillation monitoring and analysis. It is also shown that the estimated damping ratio obtained using the EMD is more accurate than a Prony analysis. The method permits an automated extraction and the characterization of a temporal modal behavior with no prior assumptions on the governing processes driving oscillations; it can be applied to a wide variety of signals found in power system oscillatory processes.

In Guo and Peng (2007) the start-up transient response of a rotor with a propagating transverse crack was investigated using the EMD method. The influence of a crack propagating ratio on an instantaneous response of the rotor – as it passes through the critical speed and subharmonic resonances – is analyzed. The one, two, and three times rotating frequency vibration components are studied when they just appear, follow the peak, and decrease during the start-up process. It is demonstrated that the EMD detects a transverse crack on the shaft, and therefore is an effective tool for the analysis of a nonlinear, unsteady transient vibration response. Some additional information on crack detection can be found in Feldman and Seibold (1999).

12.4   Conclusions

The HT-based technique enables us to estimate directly the system's instantaneous dynamic parameters (i.e., natural frequencies and damping characteristics) and also their dependence on the vibration amplitude and frequency. This direct time-domain techniques allows for a direct extraction of the linear and nonlinear system parameters from the measured time signals of input and output. The nonlinear model of the tested structures opens the way to improve the effective dynamics and monitor the health of a structure. Furthermore, with model-based procedures, it might be possible to distinguish between specific damages, such as a crack and an increasing imbalance.

The results obtained may be used to verify and validate the model under different conditions, to simulate possible solutions generated by any other input force, and to find a control scheme that provides the desired vibration response. The identification method of free and forced vibration analysis, which determines the instantaneous modal parameters, contributes to the efficient and more accurate testing of nonlinear oscillatory systems, avoiding time-consuming measurement and analysis. Sometimes a combination of different methodologies will yield an improved ability to detect damages at early stages – compared to using each approach separately.

References

Adamopoulos, P., Fong, W., and Hammond, J.K. (1988) Envelope and Instantaneous Phase Characterisation of Nonlinear System Response. Proc. of the VI Int. Modal Analysis Conf.

Agneni, A., and Balis-Crema, L. (1989). Damping measurements from truncated signals via Hilbert transform. *Mechanical Systems and Signal Processing*, **3** (1), 1–13.

Andrade, M.A., Messina, A.R., Rivera, C.A., and Olguin, D. (2004) Identi cation of instantaneous attributes of torsional shaft signals using the Hilbert transform. *IEEE Transactions on Power Systems*, **19** (3), 1422–1429.

Andronov, A.A., Vitt, A.A., and Khaikin, S.E. (1966) *Theory of Oscillators*, Pergamon Press.

Anishchenko, V.S., Astakhov, V., Neiman, A., *et al.* (2007) *Nonlinear Dynamics of Chaotic and Stochastic Systems*, Springer, Berlin, Heidelberg.

Antonino-Daviu, J.A., Riera-Guasp, M., Roger-Folch, J., and Perez, R.B. (2007) An Analytical Comparison Between DWT and Hilbert-Huang-Based Methods for the Diagnosis of Rotor Asymmetries in Induction Machines. Industry Applications Conference, 42nd IAS Annual Meeting, IEEE.

Attoh-Okine, N., Barner, K., Bentil, D., and Zhang, R. (2008) The empirical mode decomposition and the Hilbert-Huang transform. *Journal on Advances in Signal Processing*, **2008**, (Article ID 251518), pp. 1–2.

Babitsky, V.I., and Krupenin, V.L. (2001) *Vibration of Strongly Nonlinear Discontinuous Systems*, Springer.

Bedrosian, E. (1963) A product theorem for Hilbert transforms. *Proceedings of the IEEE*, **51** (5), 868–869.

Bellizzi, S., Guillemain, P., and Kronland-Martinet, R. (2001) Identi cation of coupled nonlinear modes from free vibration using time-frequency representations. *Journal of Sound and Vibration*, **243** (2), 191–213.

Bendat, J.S. (1985) *The Hilbert Transform and Applications to Correlation Measurements*, Bruel & Kjaer, Denmark.

Bernal, D., and Gunes, B. (2000) An Examination of Instantaneous Frequency as a Damage Detection Tool. 14th Engineering Mechanics Conference (EM2000), the American Society of Civil Engineers, Austin, TX.

Boashash, B. (1992) Estimating and interpreting the instantaneous frequency — Part 1: Fundamentals. *Proceedings of the IEEE*, **80** (4), 520–538.

Boller, C., Chang, F.-K., and Fujino, Y. (eds) (2009) *Encyclopedia of Structural Health Monitoring*, Wiley.

Braun, S.G. (ed.) (1986) *Mechanical Signature Analysis: Theory and Applications*, Academic Press, New York.

Braun, S., and Feldman, M. (1997) Time-frequency characteristics of non-linear systems. *Mechanical Systems and Signal Processing*, **11** (4), 611–620.

Braun, S., and Feldman, M. (2011) On the decomposition of nonstationary signals: some aspects of the EMD and HVD methods. *Mechanical Systems and Signal Processing*, **25** (5).

Brennan, M.J., Kovacic, I., Carrella, A., and Waters, T.P. (2008) On the jump-up and jump-down frequencies of the Duf ng oscillator. *Journal of Sound and Vibration*, **318**, 1250–1261.

Broman, H. (1981) The instantaneous frequency of a Gaussian signal: The one-dimensional density function. *Acoustics, Speech and Signal Processing IEEE Transactions on*, **29** (1), 108–111.

Browne, T.J., Vittal, V., Heydt, G.T., & Messina, A.R. (2009) Practical application of Hilbert transform techniques in identifying inter-area oscillations, in A.R. Messina (ed.), *Inter-area Oscillations in Power Systems*, Springer, pp. 101–125.

Bruns, J.-U., Lindner, M., and Popp, K. (2003), Identi cation of the nonlinear restoring force characteristic of a rubber mounting. *Proceedings in Applied Mathematics and Mechanics*, **2**, 270–271.

Bucher, I. (2011) Transforming and separating rotating disk vibrations using a sensor array. *Journal of Sound and Vibration*, **330** (6), 1244–1264.

Bucher, I., and Ewins, D.J. (1997) Multi-dimensional decomposition of time-varying vibration response signals in rotating machinery. *Mechanical Systems and Signal Processing*, **11** (4), 577–601.

Bucher, I., Feldman, M., Minikes, A., and Gabay, R. (2004) Real-time traveling waves and whirl decomposition, Proc. of the VIII Int. *Vibrations in Rotating Machinery*, The Institution of Mechanical Engineers.

Bunimovich, V.I. (1951) *Fluctuating Processes in Radio Receivers* (in Russian), Soviet Radio, Moscow.

Caciotta, M., Giarnetti, S., Leccese, F., and Leonowicz, Z. (2009) Detection of Short Transients and Interruptions Using the Hilbert Transform. *XIX IMEKO World Congress Fundamental and Applied Metrology*, IMEKO, Lisbon, Portugal.

Cain, G.D., Lever, K.V., and Yardim, A. (1998) Probability density functions of amplitude modulated random signals. *Electronics Letters*, **34** (16), 1560–1561.

Cantrell, C.D. (2000) *Modern Mathematical Methods for Physicists and Engineers*, Cambridge University Press.

Chen, Y., and Feng, M.Q. (2003) A technique to improve the empirical mode decomposition in the Hilbert-Huang transform. *Earthquake Engineering and Engineering Vibration*, **2** (1), 75–85.

Chen, T.-L., Que, P.-W., Zhang, Q., and Liu, Q.-K. (2005) Ultrasonic signal identi cation by empirical mode decomposition and Hilbert transform. *Review of Scientific Instruments*, **76** (8), 1–6.

Chopra, A.K. (2007) *Dynamics of Structures: Theory and Applications to Earthquake Engineering*, Pearson Prentice Hall.

Claerbout, J.F. (1976) *Fundamentals of Geophysical Data Processing*, McGraw-Hill, New York.

Cohen, L. (1989) Time-frequency distributions - a review. *Proceedings of the IEEE*, **77** (7), 941–980.

Cohen, L., and Lee, C. (1989) Standard Deviation of Instantaneous Frequency. International Conference ICASSP-89, IEEE, Glasgow, UK.

Cohen, L., and Loughlin, P. (2003) Authors' reply. *Signal Processing*, **83**, 1821–1822.

Cohen, L., Loughlin, P., and Vakman, D. (1999) On an ambiguity in the de nition of the amplitude and phase of a signal. *Signal processing*, **79**, 301–307.

Curadelli, R.O., Riera, J.D., Ambrosini, D., and Amania, M.G. (2008) Damage detection by means of structural damping identi cation. *Engineering Structures 30*, **30**, 3497–3504.

Daetig, M., and Schlurmann, T. (2004) Performance and limitations of the Hilbert-Huang transformation (HHT) with an application to irregular water waves. *Ocean Engineering*, **31**, 1783–1834.

Davidson, K.L., and Loughlin, P.J. (2000) Instantaneous spectral moments. *Journal of the Franklin Institute*, **337**, 421–436.

Davies, P., and Hammond, J.K. (1987) The Use of Envelope and Instantaneous Phase Methods for the Response of Oscillatory Nonlinear Systems to Transients. Proc. of the 5th IMAC.

Douka, E., and Hadjileontiadis, L.J. (2005) Time-frequency analysis of the free vibration response of a beam with a breathing crack. *NDT & E International*, **38** (1), 3–10.

Dragomir, S.S. (2005) Reverses of the triangle inequality in Banach spaces. *Journal of Inequalities in Pure and Applied Mathematics*, **6** (5), 77–99.

Du, Q., and Yang, S. (2007) Application of the EMD method in the vibration analysis of ball bearings. *Mechanical Systems and Signal Processing*, **21** (6), 2634–2644.

Elizalde, H., and Imregun, M. (2006) An explicit frequency response function formulation for multi-degree-of-freedom non-linear systems. *Mechanical Systems and Signal Processing*, **20**, 1867–1882.

Eret, P., and Meskell, C. (2008) A practical approach to parameter identi cation for a lightly damped, weakly nonlinear system. *Journal of Sound and Vibration*, **310**, 829–844.

Ewins, D.J. (1984) *Modal Testing: Theory, Practice and Application*, Research Studies Press, Ltd.

Fang, X., Luo, H., and Tang, J. (2008) Investigation of granular damping in transient vibrations using Hilbert transform based technique. *Journal of Vibration and Acoustics*, **130** (3), 1–11.

Fan, X., and Zuo, M.J. (2007) Gearbox fault feature extraction using Hilbert transform, S-transform, and a statistical indicator. *Journal of Testing and Evaluation*, **35** (5), 477–485.

Feeny, B.F. (2008) A complex orthogonal decomposition for wave motion analysis. *Journal of Sound and Vibration*, **310**, 77–90.

Feldman, M.S. (1985) Investigation of the natural vibrations of machine elements using the Hilbert transform. *Soviet Machine Science*, **2**, 44–47.

Feldman, M. (1991) Device and method for determination of vibration system modal parameters. Patent, 098985, Israel, Israel.

Feldman, M. (1994a) Non-linear system vibration analysis using Hilbert transform – I. Free vibration analysis method "FREEVIB". *Mechanical Systems and Signal Processing*, 119–127.

Feldman, M. (1994b) Non-linear system vibration analysis using Hilbert transform – II. Forced vibration analysis method "FORCEVIB". *Mechanical Systems and Signal Processing*, 309–318.

Feldman, M. (1997) Non-linear free vibration identi cation via the Hilbert transform. *Journal of Sound and Vibration*, **208** (3), 475–489.

Feldman, M. (2001) Hilbert transforms, in *Encyclopedia of Vibration* (eds D. Ewins, S. Braun, and S. Rao), Academic Press, New York.

Feldman, M. (2005) Time-Varying and Non-Linear Dynamical System Identi cation Using the Hilbert Transform. 20th Biennial Conference on Mechanical Vibration and Noise, ASME, Long Beach, CA.

Feldman, M. (2006) 'Time-varying vibration decomposition and analysis based on the Hilbert transform. *Journal of Sound and Vibration*, **295** (3–5), 518–530.

Feldman, M. (2007a) Considering high harmonics for identi cation of nonlinear systems by Hilbert transform. *Mechanical Systems and Signal Processing*, **21** (2), 943–958.

Feldman, M. (2007b) Identi cation of weakly nonlinearities in multiple coupled oscillators. *Journal of Sound and Vibration*, **303**, 357–370.

Feldman, M. (2008a) MATLAB® (Executable) Programs for the Hilbert Vibration Decomposition, viewed 2010, http://hitech.technion.ac.il/feldman/HT_decomposition.zip.

Feldman, M. (2008b) Theoretical analysis and comparison of the Hilbert transform decomposition methods. *Mechanical Systems and Signal Processing*, **22** (3), 509–519.

Feldman, M. (2009a) Analytical basics of the EMD: Two harmonics decomposition. *Mechanical Systems and Signal Processing*, **23** (7), 2059–2071.

Feldman, M. (2009b) Hilbert transform, envelope, instantaneous phase and frequency, in *Encyclopedia of Structural Health Monitoring* (eds C. Boller, F.-K. Chang, and Y Fujino), John Wiley & Sons Ltd.

Feldman, M. (2010) MATLAB Programs for the HT Identi cation, viewed 2010, http://hitech.technion.ac.il/feldman/HT_identi cation.zip.

Feldman, M. (2011) Hilbert transform in vibration analysis. *Mechanical Systems and Signal Processing*, **25** (3), 735–801.

Feldman, M., and Braun, S. (1993) Analysis of Typical Non-Linear Vibration Systems by Using the Hilbert Transform. Proc. of the XI Int. Modal Analysis Conf., Kissimmee, Florida.

Feldman, M., and Seibold, S. (1999) Damage diagnosis of rotors: application of Hilbert-transform and multi-hypothesis testing. *Journal of Vibration and Control*, **5**, 421–445.

Fidlin, A. (2005) *Nonlinear Oscillations in Mechanical Engineering*, Springer, Berlin, Heidelberg.

Fink, L.M. (1966) Relations between the spectrum and instantaneous frequency of a signal. *Problemy Peredachi Informatsii*, **2** (4), 26–38.

Fink, D. (1975) Coherent detection signal-to-noise. *Applied Optics*, **14** (3), 689–690.

Flandrin, P., Rilling, G., and Goncalves, P. (2004) Empirical mode decomposition as a lter bank. *IEEE Signal Processing Letters*, **11** (2), 112–114.

Franchetti, P., and Modena, C. (2009) Nonlinear damping identi cation in precast prestressed reinforced concrete beams. *Computer-Aided Civil and Infrastructure Engineering*, **24**, 577–592.

Gabor, D. (1946) Theory of communication. *Journal of the Institution of Electrical Engineers*, **93** (3), 429–457.

Gendelman, O.V., Starosvetsky, Y., and Feldman, M. (2008) Attractors of harmonically forced linear oscillator with attached nonlinear energy sink I: Description of response regimes. *Nonlinear Dynamics*, **51** (1-2), 31–46.

Gianfelici, F., Biagetti, G., Crippa, P., and Turchetti, C. (2007) Multicomponent AM-FM representations: an asymptotically exact approach. *IEEE Transactions On Audio, Speech, and Language Processing*, **15** (3), 823–837.

Giorgetta, F., Gobbi, M., and Mastinu, G. (2007) On the testing of vibration performances of road vehicle suspensions. *Experimental Mechanics*, **47** (4), 485–495.

Girolami, G., and Vakman, D. (2002) Instantaneous frequency estimation and measurement: a quasi-local method. *Measurement Science and Technology*, **13**, 909–917.

Gloth, G., and Sinapius, M. (2004) Analysis of swept-sine runs during modal identi cation. *Mechanical Systems and Signal Processing*, **18**, 1421–1441.

Goge, D., Sinapius, M., Fullekrug, U., and Link, M. (2005) Detection and description of nonlinear phenomena in experimental modal analysis via linearity plots. *International Journal of Non-Linear Mechanics*, **40**, 27–48.

Goswami, J.C., and Hoefel, A.E. (2004) Algorithms for estimating instantaneous frequency. *Signal Processing*, **84**, 1423–1427.

Gottlieb, O., and Feldman, M. (1997) Application of a Hilbert transform-based algorithm for parameter estimation of a nonlinear ocean system roll model. *Journal of Offshore Mechanics and Arctic Engineering*, **119** (4), 239–243.

Gottlieb, O., Feldman, M., and Yim, S.C.S. (1996) Parameters identi cation of nonlinear ocean mooring systems using the Hilbert transform. *Journal of offshore mechanics and Arctic engineering*, **118** (1), 29–36.

Gravier, B.M., Napal, N.J., Pelstring, J.A., *et al.* (2001) An Assessment of the Application of the Hilbert Spectrum to the Fatigue Analysis of Marine Risers. Proceedings of the Eleventh (2001) International Offshore and Polar Engineering Conf., The International Society of Offshore and Polar Engineers, Stavanger, Norway.

Grimaldi, M., and Cummins, F. (2008) Speaker identi cation using instantaneous frequencies. *IEEE Transactions on Audio, Speech, and Language Processing*, **16** (6), 1-97–1111.

Guo, D., and Peng, Z.K. (2007) Vibration analysis of a cracked rotor using Hilbert-Huang transform. *Mechanical Systems and Signal Processing*, **21**, 3030–3041.

Hahn, S.L. (1996a) *Hilbert Transforms in Signal Processing*, Artech House.

Hahn, S.L. (1996b) The Hilbert transform of the product a(t)cos(omega_ot + phi_o). *Bulletin of the Polish Academy of Sciences Technical Sciences*, **44** (1), 75–80.

Hammond, J.K. (1968) On the response of single and multidegree of freedom systems to nonstationary excitations. *Journal of Sound and Vibration*, **7**, 393–419.

Hammond, J.K., and Braun, S. (1986) Additional techniques, in *Mechanical Signature Analysis*, (ed. S. Braun), Academic Press, London.

Hartmann, W. (1998) *Signal, Sound and Sensation*, American Institute of Physics.

Herlufsen, H. (1984) *Technical Review: "Dual Channel FFT Analysis (Part II)"*, Brüel & Kjaer.

Hertz, D. (1986) Time delay estimation by combining ef cient algorithms and generalized cross-correlation methods. *IEEE Transactions on Acoustics, Speech, and Signal Processing*, **ASSP-34** (1), 1–7.

Hogan, J.A., and Lakey, J.D. (2005) Time-frequency and time-scale methods: Adaptive Decompositions, Uncertainty Principles, and Sampling, Birkhauser, Boston.

Ho, D. and Randall, R.B. (2000) Optimization of bearing diagnostic techniques using simi lated and actual bearing fault signals, *Mechanical Systems and Signal Processing*, **14** (5), 763–788.

Huang, D. (2003) Practical implementation of the Hilbert-Huang transform algorithm. *Acta Oceanologica Sinica*, **25** (1), 1–11.

Huang, N., and Shen, S. (eds.) (2005) Hilbert-Huang transform and its applications, in *Interdisciplinary Mathematical Sciences (Interdisciplinary Mathematical Sciences)*, World Scienti c.

Huang, N.E., Shen, Z., and Long, S.R. (1999) New view of nonlinear water waves: the Hilbert spectrum. *Annual Review of Fluid Mechanics*, **31**, 417–457.

Huang, N.E., Shen, Z., Long, S.R., *et al.* (1998) The empirical mode decomposition and the Hilbert spectrum for nonlinear and non-stationary time series analysis. *Proceedings of the Royal Society of London. Series A.*, **454** (1971), 903–995.

Huang, N.E., and Wu, Z. (2008) A review on Hilbert-Huang transform: Method and its applications to geophysical studies. *Reviews of Geophysics*, **46**, pp. 1–23.

Huang, N., Wu, Z., Long, SR., *et al.* (2009) On instantaneous frequency. *Advances in Adaptive Data Analysis*, **1** (2), 177–229.

Hu, H.F., Staszewski, W.J., Hu, N.Q., *et al.* (2010) Crack detection using nonlinear acoustics and piezoceramic transducers - instantaneous amplitude and frequency analysis. *Smart Materials and Structures*, **19**, 1–10.

Hutchinson, J.W., and Wu, T.Y. (1996) *Advances in Applied Mechanics*, Academic Press, Inc.

Inaudi, J.A., and Kelly, J.M. (1995) Linear hysteretic damping and the Hilbert transform. *Journal of Engineering Mechanics*, **121**, 626–632.

Inman, D.J. (1994) *Engineering Vibration*, Prentice Hall.

Johansson, M. (1999) The Hilbert Transform, viewed 2010, http://w3.msi.vxu.se/exarb/mj_ex.pdf.

Jones, J. (2009) Embedded algorithms within an FPGA to classify nonlinear single-degree-of-freedom systems. *Sensors Journal, IEEE*, **9** (11), 1486–1493.

Kaiser, J.F. (1990) On Teager's Energy Algorithm and its Generalization to Continuous Signals. Proc. IEEE Digital Signal Process. Workshop, New Paltz, NY.

Kendig, R.P. (1997) Algorithm to Count Fatigue Cycles. Proceedings-National-Conference-on-Noise-Control-Engineering, Inst of Noise Control Engineering, NY.

Kerschen, G., Vakakis, A.F., Lee, Y.S., *et al.* (2008) Toward a fundamental understanding of the Hilbert-Huang transform in nonlinear structural dynamics. *Journal of Vibration and Control*, **14** (1-2), 77–105.

Kerschen, G., Worden, K., Vakakis, A.F., and Golinval, J.-C. (2006) Past, present and future of nonlinear system identi cation in structural dynamics. *Mechanical Systems and Signal Processing*, **20** (3), 505–592.

Kijewski-Correa, T., and Kareem, A. (2006) Ef cacy of Hilbert and Wavelet transforms for time-frequency analysis. *Journal of Engineering Mechanics*, **132** (10), 1037–1049.

King, F.W. (2009) *Hilbert Transforms: (Encyclopedia of Mathematics and Its Applications)*, Cambridge University Press, Cambridge.

Kizhner, S., Blank, K., Flatley, T., *et al.* (2006) On Certain Theoretical Developments Underlying the Hilbert-Huang Transform. Aerospace Conference, IEEE.

Korpel, A. (1982) Gabor: frequency, time, and memory. *Applied Optics*, **21** (20), 3624–3632.

Kovesi, P. (1999) Image features from phase congruency. *A Journal of Computer Vision Research*, **1** (3), 1–26.

Kultyshev, A.S. (1990) Envelope, frequency, and phase representation in terms of a family of oscillations. *Measurement Techniques*, **33** (8), 839–842.

Laila, D.S., Larsson, M., Pal, B.C., and Korba, P. (2009) Nonlinear Damping Computation and Envelope Detection using Hilbert Transform and its Application to Power Systems Wide Area Monitoring. IEEE Power & Energy Society General Meeting, PES '09.

Laila, D.S., Messina, A.R., and Pal, B.C. (2009) A re ned Hilbert-Huang transform with applications to inter area oscillation monitoring. *IEEE Transactions On Power Systems*, **24** (2), 610–620.

Lai, Y.-C., and Ye, N. (2003) Recent developments in chaotic time series analysis. *International Journal of Bifurcation and Chaos*, **13** (6), 1383–1422.

Lang, Z.Q., and Billings, S.A. (2005) Energy transfer properties of nonlinear systems in the frequency domain. *International Journal of Control*, **78**, 354–362.

Lee, C.W., and Han, Y.-S. (1998) Use of Directional Wigner Distribution for Identi cation of the Instantaneous Whirling Orbit in Rotating Machinery. The Seventh International Symposium on Transport Phenomena and Dynamics of Rotating Machinery.

Lee, Y.S., Vakakis, A.F., McFarland, D.M., and Bergman, L.A. (2010) Non-linear system identi cation of the dynamics of aeroelastic instability suppression based on targeted energy transfers. *The Aeronautical Journal*, **114** (1152), 61–82.

Liang, J.. Chaudhuri, S.R., and Shinozuka, M. (2007) Simulation of nonstationary stochastic processes by spectral representation. *Journal of Engineering Mechanics*, **133** (6), 616–627.

Li, H., Deng, X., and Dai, H. (2007) Structural damage detection using the combination method of EMD and wavelet analysis. *Mechanical Systems and Signal Processing*, **21** (1), 298–306.

Lin, S., Yang, J., and Zhou, L. (2005) Damage identi cation of a benchmark building for structural health monitoring. *Smart Materials and Structures*, **14**, 162–169.

Liu, S., Messina, A.R., and Vittal, V. (2004) Characterization of Nonlinear Modal Interaction using Normal Forms and Hilbert Analysis. Proc. of 2004 IEEE PES Power Systems Conference & Exposition New York, New York.

Liu, B., Riemenschneider, S., and Xu, Y. (2006) Gearbox fault diagnosis using empirical mode decomposition and Hilbert spectrum. *Mechanical Systems and Signal Processing*, **20**, 718–734.

Loughlin, P.J., and Tacer, B. (1997) Comments on the interpretation of instantaneous frequency. *IEEE Signal Processing Letters*, **4** (5), 123–125.

Luo, H., Fang, X., and Ertas, B. (2009) Hilbert transform and its engineering applications. *AIAA Journal*, **47** (4), 923–932.

Lyons, R. (2000) Quadrature Signals: Complex, But Not Complicated, viewed. (2010) http://www.elektronikschule.de/~krausg/DSP/Dsp_CD/Quadrature_Signal_Tutorial/quadsig.htm.

Lyons, R.G. (2004) *Understanding Digital Signal Processing*, Prentice Hall.

MacDonald, J.R., and Brachman, M.K. (1956) Linear-system integral transform relations. *Reviews of Modern Physics*, **28** (4), 393–422.

Mai, E.C., Sugeng, Y.P., Pei, J.S., *et al.* (2008) Design and Testing of Small Timber Models to Demonstrate Nonlinear Dynamics. International Modal Analysis Conference (IMAC XXVI), Orlando, FL.

Makris, N. (1999) Frequency-independent dissipation and causality, in *Computational Stochastic Mechanics* (ed. P. Spanos), Balkema, Rotterdam, pp. 435–442.

Manske, R.A. (1968) Computer simulation of narrowband systems. *IEEE Transactions on Computers*, **C-17** (4), 301–308.

Mercier, M.J., Garnier, N.B., and Dauxois, T. (2008) Re ection and diffraction of internal waves analyzed with the Hilbert transform. *Physics of Fluids*, **20** (8), 086601–086610.

Messina, A.R. (ed.) (2009) *Inter-area Oscillations in Power Systems: A Nonlinear and Nonstationary Perspective*, Springer US.

Migulin, V.V., Medvedev, V.Y., Mustel, E.R., and Parigin, V.N. (1983) *Basic Theory of Oscillations*, Mir Publishers, Moscow.

Minikes, A.. Gabay, R.. Bucher, I, and Feldman, M. (2005) On the sensing and tuning of progressive structural vibration waves. *IEEE Transactions on Ultrasonics, Ferroelectrics and Frequency Control*, **52** (9), 1565–1576.

Mitra, S.K., and Kaiser, J.F. (1993) *Handbook for Digital Signal Processing*, John Wiley & Sons.

Nayfeh, A.H. (1979) *Nonlinear Oscillations*, Wiley-Interscience.

Nho, W., and Loughlin, P.J. (1999) When is instantaneous frequency the average frequency at each time? *IEEE Signal Processing Letters*, **6** (4), 78–80.

Pai, P.F. (2007) Nonlinear vibration characterization by signal decomposition. *Journal of Sound and Vibration*, **307**, 527–544.

Pai, P.F., and Hu, J. (2006) Nonlinear Vibration Characterization by Signal Decomposition. IMAC-XXIV Conference and Exposition on Structural Dynamics, St. Louis, Missouri.

Pai, P., Huang, L., Hu, J., and Langewisch, D. (2008) Time-frequency method for nonlinear system identi cation and damage detection. *Structural Health Monitoring-an International Journal*, **7** (2), 103–127.

Pai, P.F., and Palazotto, A.N. (2008) Detection and identi cation of nonlinearities by amplitude and frequency modulation analysis. *Mechanical Systems and Signal Processing*, **22**, 1107–1132.

Pandey, J.N. (1996) *The Hilbert Transform of Schwartz Distributions and Applications*, John Wiley & Sons, New York.

Pei, J.-S., and Piyawat, K. (2008) Deterministic excitation forces for simulation and identi - cation of nonlinear hysteretic SDOF systems. *Journal of Engineering Mechanics*, **134** (1), 35–48.

Perry, P., and Brazil, T. (1997) Hilbert-transform-derived relative group delay. *IEEE Transactions on Microwave Theory and Techniques*, **45** (8), 1214–1225.

Pines, D.J., and Salvino, L.W. (2006) Structural damage detection using empirical mode decomposition and HHT. *Journal of Sound and Vibration*, **294**, 97–124.

Plakhtienko, N.P. (2000) Methods of identi cation of nonlinear mechanical vibrating systems. *International Applied Mechanics*, **36** (12).

Poon, C.W., and Chang, C.C. (2007) Identi cation of nonlinear elastic structures using empirical mode decomposition and nonlinear normal modes. *Smart Structures and Systems*, **3** (2), 423–437.

Potamianos, A., and Maragos, P. (1994) A comparison of the energy operator and the Hilbert transform approach to signal and speech demodulation. *Signal Processing*, **37**, 95–120.

Putland, G.R., and Boashash, B. (2000) Can a Signal be Both Monocomponent and Multicomponent? Third Australasian Workshop on Signal Processing Applications (WoSPA 2000), Brisbane.

Qian, T. (2006) Mono-components for decomposition of signals. *Mathematical Methods in the Applied Sciences*, **29** (10), 1187–1198.

Quek, S.T., Tua, P.S., and Wang, Q. (2003) Detecting anomalies in beams and plate based on the Hilbert-Huang transform of real signals. *Smart Materials and Structures*, **12**, 447–460.

Randall, R.B. (1986) Hilbert Transform Techniques in Machine Diagnostics, IFToMM International Conference on Rotordynamics, Tokyo.

Raman, S., Yim, S.C.S., and Palo, P.A. (2005) Nonlinear model for sub- and superharmonic motions of a MDOF moored structure, Part 1-system identi cation. *Journal of Offshore Mechanics and Arctic Engineering*, **127**, 283–290.

Rilling, G., and Flandrin, P. (2008) One or two frequencies? the empirical mode decomposition answers. *IEEE Transactions on Signal Processing*, **56** (1), 85–95.

Rosenblum, M.G. (1993) A characteristic frequency of chaotic dynamical system. *Chaos, Solutions & Fractals*, **3** (6), 617–626.

Rosenblum, M.G., Pikovsky, A.S., and Kurths, J. (1996) Phase synchronization of chaotic oscillators. *Physical Review Letters*, **76**, 1804–1807.

Ruzzene, M. (2007) Frequency-wavenumber domain ltering for improved damage visualization. *Smart Materials and Structures*, **16**, 2116–2129.

Salvino, L.W. (2000) Empirical Mode Analysis of Structural Response and Damping. International Modal Analysis Conference, Proc. IMAC-XVIII: A Conference on Structural Dynamics, Society for Experimental Mechanics, Bethel, Conn., San Antonio.

Salvino, L.W., Pines, D.J., Todd, M., and Nichols, J. (2005) EMD and instantaneous phase detection of structural damage, in *Hilbert-Huang Transform: Introduction and Applications* (ed. N. Huang and S. Chen), World Scienti c Publishing.

Savitzky, A., and Golay, M.J.E. (1964) Smoothing and differentiation of data by simpli ed least-squares procedures. *Analytical Chemistry*, **36** (8), 1627–1639.

Schreier, P.J., and Scharf, L.L. (2010) Statistical *Signal Processing of Complex-Valued Data: The Theory of Improper and Noncircular Signals*, Cambridge University Press.

Sharpley, R.C., and Vatchev, V. (2006) Analysis of the intrinsic mode functions. *Constructive Approximation*, **24** (1), 17–47.

Shi, Z.Y., and Law, S.S. (2007) Identi cation of linear time-varying dynamical systems using Hilbert transform and empirical mode decomposition method. *Journal of Applied Mechanics*, **74** (2), 223–230.

Shin, K., and Hammond, J. (2008) *Fundamentals of Signal Processing for Sound and Vibration Engineers*, John Wiley & Sons, Ltd.

Simon, M., and Tomlinson, G.R. (1984) Use of the Hilbert transform in modal analysis of linear and non-linear structures. *Journal of Sound and Vibration*, **96** (4), 421–436.

Smith, C.B., and Wereley, N.M. (1999) Nonlinear damping identi cation from transient data. *AIAA Journal*, **37** (12), 2–19.

Starosvetsky, Y., and Gendelman, O.V. (2008) Attractors of harmonically forced linear oscillator with attached nonlinear energy sink. II: Optimization of a nonlinear vibration absorber. *Nonlinear Dynamics*, **51** (1-2), 47–57.

Starosvetsky, Y., and Gendelman, O.V. (2010) Interaction of nonlinear energy sink with a two degrees of freedom linear system: Internal resonance. *Journal of Sound and Vibration*, **329** (10), 1836–1852.

Suzuki, H., Ma, F., Izumi, H., *et al.* (2006) Instantaneous frequencies of signals obtained by the analytic signal method. *Acoustical Science and Technology*, **27** (3), 163–170.

Ta, M.-N., and Lardies, J. (2006) Identi cation of weak nonlinearities on damping and stiffness by the continuous wavelet transform. *Journal of Sound and Vibration*, **293**, 16–37.

Teager, H.M. (1980) Some observations on oral air ow during phonation. *IEEE Trans. Acoust. Speech Signal Process*, **ASSP-28**, 599–601.

Therrien, C. (2002) The Lee-Wiener Legacy. A history of the statistical theory of communication. *IEEE Signal Processing Magazine*, **19** (6), 33–44.

Thomas, T.G., and Sekhar, S.C. (2005) *Communication Theory*, Tata Mcgraw Hill Publihsers.

Thrane, N., Wismer, J., Konstantin-Hansen, H., and Gade, S. (1984) Application Note. Practical use of the Hilbert transform, Technical Review No. 3, viewed 2010, http://www.bksv.com/pdf/Bo0437.pdf.

Titchmarsh, E.C. (1948) *Introduction to the Theory of Fourier Integrals*, Clarendon Press.

Tjahjowidodo, T., Al-Bender, F., and Van Brussel, H. (2007) Identi cation of Backlash in Mechanical Systems Experimental dynamic identi cation of backlash using skeleton methods. *Mechanical Systems and Signal Processing*, **21**, 959–972.

Tomlinson, G.R. (1987) Developments in the use of the Hilbert transform for detecting and quantifying non-linearity associated with frequency response functions. *Mechanical Systems and Signal Processing*, **1** (2), 151–171.

Tomlinson, G.R., and Ahmed, I. (1987) Hilbert transform procedures for detecting and quantifying non-linearity in modal testing. *Meccanica*, **22** (3), 123–132.

Turner, C. (2009) An Ef cient Analytic Signal Generator. *IEEE Signal Processing Magazine*, **26** (3), 91–94.

Vainshtein, L.A., and Vakman, D.E. (1983) *Frequency Separation in the Theory of Vibration and Waves* (in Russian), Nauka, Moscow.

Vakman, D. (1996) On the analytic signal, the Teager-Kaiser energy algorithm, and other methods for de ning amplitude and frequency. *IEEE Transactions on Signal Processing*, **44** (4), 792–797.

Vakman, D. (1998) *Signals, Oscillations, and Waves*, Artech House, Boston, London.

Vakman, D. (2000) New high precision frequency measurement. *Measurement Science and Technology*, **11** (10), 1493–1497.

Vakman, D.E., and Vainshtein, L.A. (1977) Amplitude, phase, frequency - fundamental concepts of Oscillation theory. *Soviet Physics - Uspekhi*, **20** (12), 1002–1016.

Veltcheva, A., Cavaco, P., and Soares, C.G. (2003) Comparison of methods for calculation of the wave envelope. *Ocean Engineering*, **30**, 937–948.

Wallace, D.A., and Darlow, M.S. (1988) Hilbert transform techniques for measurement of transient gear. *Mechanical Systems and Signal Processing*, **2** (2), 187–194.

Wang, L., Zhang, J., Wang, C., and Hu, S. (2003) Time-frequency analysis of nonlinear systems: the skeleton linear model and the skeleton curves. *Transactions of the ASME*, **125**, 170–177.

Weaver, W., Timoshenko, S.P., and Young, D.H. (1990) *Vibration Problems in Engineering*, John Wiley.

Wei, D., and Bovik, A.C. (1998) On the instantaneous frequencies of multicomponent AM-FM signals. *IEEE Signal Processing Letters*, **5** (4), 84–86.

Wetula, A. (2008) A Hilbert Transform Based Algorithm for Detection of a Complex Envelope of a Power Grid Signals - an Implementation. *Electrical Power Quality & Utilization - Journal + Magazine*, **XIV** (2), 13–18.

Whitaker, J.C. (2005) *The Electronics Handbook*, CRC Press.

Worden, K., and Manson, G. (1998) Random vibrations of a Duf ng oscillator using the Volterra series. *Journal of Sound and Vibration*, **217** (4), 781–789.

Worden, K., and Tomlinson, G.R. (2001) *Nonlinearity in Structural Dynamics: Detection, Identification, and Modelling*, IOP Publishing Ltd, Bristol, United Kingdom.

Wu, Z. (2008) The module that performs EMD/EEMD, viewed December 2010, http://rcada.ncu.edu.tw/research1_clip_program.htm.

Wu, Z., and Huang, N.E. (2004) A study of the characteristics of white noise using the empirical mode decomposition method. *Proceedings of the Royal Society A*, **460**, 1597–1611.

Wu, Z., and Huang, N.E. (2009) Ensemble empirical mode decomposition: a noise-assisted data analysis method. *Advances in Adaptive Data Analysis*, **1** (1), 1–41.

Wu, F., and Qu, L. (2008) An improved method for restraining the end effect in empirical mode decomposition and its applications to the fault diagnosis of large rotating machinery. *Journal of Sound and Vibration*, **314** (3-5), 586–602.

Yang, J.-N. (1972) Simulation of random envelope processes. *Journal of Sound and Vibration*, **21** (1), 73–85.

Yang, J.N., Kagoo, P.K., and Lei, Y. (2000) Parametric Identi cation of Tall Buildings Using Ambient Wind Vibration Data. The 8th ASCE Specialty Conference on Probabilistic Mechanics and Structural Reliability, University of Notre Dame, Notre Dame, Indiana.

Yang, L., Lei, Y., Lin, S., and Huang, N. (2004) Hilbert-Huang based approach for structural damage detection. *Journal of Engineering Mechanics*, **130** (1), 85–95.

Yang, J.N., Lei, Y., Pan, S., and Huang, N. (2003a) Identi cation of linear structures based on Hilbert-Huang transform. Part 1: normal modes. *Earthquake Engineering and Structural Dynamics*, **32** (9), 1443–1467.

Yang, J.N., Lei, Y., Pan, S., and Huang, N. (2003b) Identi cation of linear structures based on Hilbert-Huang transform. Part 2: complex modes. *Earthquake Engineering and Structural Dynamics*, **32** (10), 1533–1554.

Yang, B., and Suh, C.S. (2004a) Interpretation of crack-induced rotor non-linear response using instantaneous frequency. *Mechanical Systems and Signal Processing*, **18**, 491–513.

Yang, B., and Suh, C.S. (2004b) On the characteristics of bifurcation and nonlinear dynamic response. *Journal of Vibration and Acoustics*, **126** (4), 574–579.

Yu, D., Cheng, J., and Yang, Y. (2005) Application of EMD method and Hilbert spectrum to the fault diagnosis of roller bearings. *Mechanical Systems and Signal Processing*, **19**, 259–270.

Index

Hilbert Transform Applications in Mechanical Vibration, First Edition. Michael Feldman.
© 2011 John Wiley & Sons, Ltd. Published 2011 by John Wiley & Sons, Ltd.

Printed and bound by CPI Group (UK) Ltd, Croydon, CR0 4YY

16/04/2025

14658544-0006